MOTION CONTROL

Multi-Faceted Movement in Space, Time and Neurological Impairment

MOTION CONTROL

Multi-Faceted Movement in Space, Time and Neurological Impairment

Yoram Baram

Technion — Israel Institute of Technology, Israel

World Scientific

NEW JERSEY • LONDON • SINGAPORE • BEIJING • SHANGHAI • HONG KONG • TAIPEI • CHENNAI • TOKYO

Published by

World Scientific Publishing Co. Pte. Ltd.
5 Toh Tuck Link, Singapore 596224
USA office: 27 Warren Street, Suite 401-402, Hackensack, NJ 07601
UK office: 57 Shelton Street, Covent Garden, London WC2H 9HE

Library of Congress Cataloging-in-Publication Data
Names: Baram, Yoram, author.
Title: Motion control : multi-faceted movement in space, time and neurological impairment / Yoram Baram.
Description: New Jersey : World Scientific, [2023] | Includes bibliographical references and index.
Identifiers: LCCN 2023001539 | ISBN 9789811269578 (hardcover) | ISBN 9789811269585 (ebook)
Subjects: LCSH: Biological control systems--Mathematical models. | Feedback control systems--Mathematical models. | Motion control devices. | Nervous system--Diseases--Patients--Rehabilitation--Technological innovations.
Classification: LCC QH508 .B26 2023 | DDC 612/.022--dc23/eng/20230313
LC record available at https://lccn.loc.gov/2023001539

British Library Cataloguing-in-Publication Data
A catalogue record for this book is available from the British Library.

For any available supplementary material, please visit
https://www.worldscientific.com/worldscibooks/10.1142/13240#t=suppl

Typeset by Stallion Press
Email: enquiries@stallionpress.com

I am deeply indebted to Ophira, Tami, Orli, Amir, Ur, Maya, Lotem, Sara, Ya'ar and Arava for being there

"*Motion Control* is composed of modules presenting, respectively, theories of stochastic dynamical system modelling; feedback control theory; and the author's work with medical collaborators on the analytic and experimental modelling of human gait problems together with their development of a virtual reality feedback device for gait stabilization, in particular for those suffering from Parkinson's Disease. *Motion Control* joins the author's preceding monograph, *The Subcritical Brain*, to make a valuable contribution to the scientific and clinical area of human cybernetics."

—**Peter E. Caines**, FRSC, Distinguished James McGill Professor, Macdonald Chair in the Department of Electrical and Computer Engineering, McGill University

"From the pioneer on closed-loop virtual feedback for the treatment of Parkinson's-related freezing of gait comes the definitive treatise on the fascinating theme of motor control. Professor Yoram Baram summons the best of probability and statistics to create a coherent framework for the understanding and modulation of critical motor behaviors, such as gait. Not for the faint of heart, this encyclopedic tour de force includes all nuts and bolts behind each argument from one of the most sophisticated computer scientists of our era."

—**Alberto J. Espay**, MD, MSc, FAAN, Professor of Neurology, Director and Endowed Chair, Gardner Family Center for Parkinson's Disease and Movement Disorders, University of Cincinnati Academic Health Center

"In his book *Motion Control*, Professor Yoram Baram presents a detailed account of modern control theory, from its mathematical foundations in linear algebra, probability, stochastic processes and information theory, through its multi-faceted realization in highly advanced engineering systems, to its seemingly unexpected realization in a virtual reality apparatus, enabling consistent and safe movement under neurological impairment. Professor Baram's work has unveiled the very details of natural and artificial control processes, so as to completely understand them and put them to work. The end results, manifested by both solid mathematics and medical achievements, is unquestionably rewarding."

—**Uri Shaked**, Professor, Tel Aviv University, Fellow of the IEEE, Fellow of the Institute of Mathematics and its Applications (IMA), Winner of the 2017 Israel Prize for Engineering Research

"I was impressed with the thoroughness and the rigor of the presentation. The topic of the book is of great importance to the many people who suffer from motion disabilities."

—**Thomas Kailath**, Hitachi America Professor of Engineering, Emeritus, Stanford University, Member: US National Academy of Engineering, US National Academy of Sciences, Recipient of the IEEE Medal of Honor and the US National Medal of Science

Prologue

As a beginning PhD student, enthusiastically immersed in the study of advanced control theory, I was called to participate in an MIT-NASA project aimed at designing a so-called Fly-by-Wire aircraft control system for large-scale flight envelopes. Some 30 years later, by sheer coincidence, I came across a truly amazing discovery by a neurologist, who, 50 years earlier, found that people inflicted with Parkinson's disease improve their impaired gait when walking perpendicularly to equally spaced line markings drawn on the floor. My earlier familiarity with the theory of feedback control, and the engineering support I received from skillful associates, have led to the development of a portable device, which provided the patient with a virtual earth-stationary tiled floor. In contrast to an open-loop, constantly moving perpendicular line sequence suggested by others, the virtual tiled floor moved in complete coordination with the patient's opposite movement, providing closed-loop feedback motion control. This approach, quickly accepted by a few leading neurologists, but not immediately understood by others, proved, in time, to be a game changer in the design of technological support for neurologically impaired patients. Over the years, collaboration with an increasingly large number of leading neurologists worldwide has resulted in highly successful clinical tests. The initial application of the device to Parkinson's patients was translated to the treatment of patients inflicted with other neurological movement impairments as well. Such disorders have ranged from cerebral palsy to multiple sclerosis, from Parkinson's disease to stroke and senile gait, from the very young to the very old neurologically impaired patients.

My interest in the mathematics of cortical behavior has resulted in a recent book, "The Subcritical Brain" (Baram, 2021). Systematically connecting cortical processes with a variety of advanced mathematical disciplines, from nonlinear dynamics to random graphs, prime numbers and quantum computation, that book's message was that it takes a blend of advanced mathematical concepts to put together a theory of cortical operation. The mathematical disciplines addressed in the present book, namely, probability theory, statistics and information theory, yield measures of accuracy, stability and efficiency in the selection of the corresponding motion control models. It is shown that the mathematical principles of control theory are relevant not only in normal cortical operation, such as target reaching and obstacle avoidance, but, with certain technological adaptation, in overcoming neurological impairment.

I am indebted to my colleagues for their help and inspiration throughout my theoretical and experimental efforts: Judith Aharon-Peretz, Amir Baram, Yair Be'ery, Peter Caines, Doron Eidelman, Alberto Espay, Thomas Kailath, Ruben Lenger, Ariel Miller, Dario Ringach, Lior Ron, Virginia de Sa, Uri Shaked, Yahalomit Simionovici and Priya Velu. Many more collaborators and authors are referenced throughout this book.

Yoram Baram
Haifa, 2023

Contents

Part III: From Natural to Virtual Model Selection in Biological Motion Control

Part IV: Gait Entrainment by Sensory Feedback in the Neurologically Impaired

Chapter 1

Introduction

1.1 Historical account

What seems to have been the first formally recognized control system (Maxwell, 1868), was followed by a slow transition from strict engineering applications, allowing for open-loop and closed-loop designs alike, to advanced mathematical analysis of dynamical systems. A significant turning point from so-called "classical" to so-called "modern" control theory is associated with the 1960's turn to the state-space model. The built-in closed-loop characteristics of this model have established not only highly formal algebraic representation of systems dynamics, but also such new notions as "observability" and "controllability". These, define, respectively, the abilities of the internal states of a system to be determined by its external outputs and their abilities to move around in their entire configuration space (Kalman, 1960a). Moreover, such understanding has largely motivated a fresh new look at a previously established mathematical discipline, namely, probability theory, and at a closely related, yet, more recent discipline, namely, information theory.

Parameter estimation techniques have been studied since the introduction of the maximum *a posteriori* probability (MAP) and the *least squares* (LS) criteria by Gauss (1809) and Laplace (1820) and their later studies by Edgeworth (1908). Fisher (1922) proposed the *maximum likelihood* (ML) estimate, which has since gained considerable popularity due to its intuitive appeal and its asymptotic properties (Lecam, 1953). The consistency (convergence to a final constant value) of ML estimates for sequences of independent and identically distributed (i.i.d.) observations was proved by Cramer (1946), who assumed differentiability to 4th order of the probability density function involved. Differentiability assumptions were dispensed with in proofs by Doob (1934) and Wald (1949). The main tool in proving consistency for i.i.d. observations is, naturally, the strong law of large numbers, first proved by Bernoulli (1713). Roussas (1965) proved the consistency of the ML estimate for the case of ergodic Markov observation sequences, employing the ergodic theorem (Birkhoff, 1931). The mean-square (MS) convergence of LS estimates given i.i.d. observations was considered by Liporace (1971), who showed, via the multiplication rule for independent random variable, that the MS error of these estimates is exponentially diminishing. In the case where the true parameter is not included in the parameter set, the estimates were shown to converge to a parameter in the set, which is most similar to the true parameter. Caines (1975a; 1988) applied the submartingale property of sequences of ML ratios on finite parameter sets to prove the consistency of the ML estimates on such sets for a general class of observation sequences, satisfying a certain probabilistic condition. Baram (1976) extended Caines' results to Bayesian estimates, which were shown to be consistent almost everywhere (a.e.) and in the mean square (m.s.), and showed that Caine's condition applies to stationary Gaussian linear systems.

The identification of linear dynamical systems employing parameter estimation techniques has been studied intensively. However, several consistency proofs that have appeared in the early literature have overlooked the fact that, for consistent estimation on compact parameter sets, uniform convergence of the associated probability densities on the parameter set is necessary, while pointwise convergence only provides consistency for finite parameter sets. Correct consistency proofs have appeared in the literature in more recent years. Caines and Rissanen (1974) (also see (Rissanen and Caines, 1974; Caine, 1988)) proved the consistency of ML estimates for autoregressive and moving average (ARMA) observation sequences. Ljung proved the consistency of a general class of stochastic approximation techniques (1974a) and the consistency of a class of prediction error techniques (1974b) (see also (Ljung, 1975)).

Caines (1975b; 1988) proved consistency for stationary processes of a more general class of prediction error techniques, which includes the ML technique for the case of stationary Gaussian observation sequences. The topological requirements specified by Caines (1975b; 1988) reduce in the finite parameter set case to a requirement that there exists a 1-to-1 correspondence between the parameter set and the set of system's impulse responses, corresponding to the system's innovations representation. Similar conditions were suggested by Tse and Weinert (1975) (see also (Tse, 1976)) and Hawkes and Moore (1976) (see also (Moore and Hawkes, 1974)), who considered the convergence of Bayesian estimates on finite sets of stationary Gaussian linear systems. The conditions suggested by Baram (1976) are uniqueness conditions on the output statistics associated with the different models in the model set.

Information methods have been suggested by several authors for the solution of estimation and control-related problems such as hypothesis testing, signal selection, and model identification. Kullback's information measure (1959) has proved to be useful in the analysis of parameter estimation and model identification techniques. Akaike (1972; 1974a; 1974b) has related Kullback's information to certain versions of the ML criterion. Kullback's information measure was employed by Liporace (1971) and, following Liporace, by Hawkes and Moore (1976) in their studies of parameter estimates given independently identically distributed (i.i.d.) and stationary Gaussian observations. In this book, following Baram (1976), we employ an information measure, which proves to possess the valuable property of being a norm on the parameter space, which is lacked by Kullback's information measure. The "Baram Proximity Measure" was discussed at length in a keynote lecture presented by the late Prof. Michael Athans of MIT, addressing issues of robust adaptive feedback control at the 2006 IFAC World Congress.

Turning to a medical application of feedback control addressed in a later part of this book, a seminal study (Martin, 1967) and subsequent clinical tests (Bagley *et al.*, 1991; Morris *et al.*, 1994a; 1994b; Azulay *et al.*, 1999) have shown significant improvement in gait parameters of patients with Parkinson's disease (PD) walking over transverse lines drawn on the ground. Early attempts to generate such visual cues by virtual means have resulted in open-loop systems, imposing, by a constantly moving visual cue (Prothero, 1993; Weghorst *et al.*, 1994; Riess and Weghorst, 1995), or a geometrically patterned treadmill (Hanakawa *et al.*, 1999), a constant walking speed on the patient. A comparison of open-loop visual cueing by virtual means to transverse line markings on the ground (Griffin *et al.*, 2011) has found the first to have a marginal effect and the second to have a significant positive effect on gait parameters in PD patients. Subject-stationary visual cues (Lewis *et al.*, 2000) represent another unnatural imposition, where the image moves in the same direction as the subject, which stands in contrast to the opposite motion of an earth-stationary image. External open-loop auditory entrainment of gait, where a rhythmic sound is imposed on the patient in a metronome-like fashion, has been studied (Kritikos *et al.*, 1995; Thaut *et al.*, 1996; McIntosh *et al.*, 1997; Howe *et al.*, 2003; Nieuwboer *et al.*, 2007), and reasoned on neurobiological grounds (Thaut *et al.*, 2014). Yet, a need for constant vigilance and attention strategies to prevent reversion to impaired gait patterns caused by repetitive stimuli has been noted (Morris *et al.*, 1996). It has been further suggested that external rhythmic entrainment instigates dopamine reward (Miendlarzewska and Trost , 2013; Salimpoor *et al.*, 2015), seemingly supporting open-loop intervention. However, open-loop control is known to be inherently unstable, with potentially disastrous consequences due to error accumulation (Kuo, 1981). In sharp contrast, closed-loop feedback systems, when correctly designed, can regulate and stabilize otherwise unstable dynamics (Kuo, 1981). Analysis has shown that walking over earth-stationary visual markings constitutes a closed-loop feedback control system, which stabilizes and regulates gait (Baram, 1999). Portable virtual realization of autonomous gait entrainment by sensory feedback, facilitated by the gradual miniaturization of sensing, computation and display technologies, has led to the development of a closed-loop sensory feedback device for gait improvement in the neurologically impaired (Baram, 2004).

1.2 Book outline

This book is aimed at connecting mathematical and practical aspects of motion, directing attention to the various control and estimation facets associated with single and with multiple models. Following the introductory historical account (Chapter 1), mathematical preliminaries (Chapter 2) are provided so as to establish the main concepts, namely, probability, statistics and information, underlying the analytic aspects of this book. Beyond these preliminaries, the book is divided into four parts. The first part of the book (Chapters 3–7) deals with optimality under size minimality of the single linear control model. Specifically, Chapter 3 addresses the issue of optimal linear state estimation. Chapter 4 addresses the issues of estimability and regulability, which are shown to be dual properties of linear control systems. Chapter 5 addresses the issue of minimal-order linear state estimation under performance constraints. Chapter 6 addresses the issue of minimal-order linear model identification. Chapter 7 addresses minimal-order predictability of future states. The second part of the book (Chapters 8–12) deals with the optimality of multi-modal estimation and control. Chapter 8 addresses the issue of distance measures between multiple general models. Chapter 9 addresses the issue of convergence in general stochastic model sets to the true model and to the nearest model in the absence of the true model, while Chapter 10 does the same for sets of linear models. A min-max proximity approach is presented in Chapter 11, while an example of multimodal aircraft control is presented in Chapter 12. The third part of the book (Chapters 13 and 14) presents the transition from naturally nonlinear control of movement associated with obstacle avoidance and object targeting to virtually linear control of movement in the neurologically impaired. The latter chapter (14) brings into sight a portable virtual reality device which serves the purpose of helping patients with neurological movement impairment, also addressing the role of dopamine reword and reward-seeking in such patients. The fourth and final part of the book (Chapters 15–22) addresses the application of virtual sensory feedback in walking with specific neurological impairments.

While the clinical studies reported were all based on a single model paradigm, a later reflection reveals that, given the variety of the neurological symptoms associated with the relevant disorders, a multi-modal approach is applicable in the treatment of neurological disorders. Chapter 15 presents the results of a seminal clinical study performed on Parkinson's disease (PD) patients, demonstrating their gait improvement resulting from closed-loop virtual visual feedback, as well as subsequent EEG recording. Clinical tests of PD patients employing auditory feedback are reported in Chapter 16, while the results achieved by such patients in at-home training with audio-visual feedback are reported in Chapter 17. Clinical tests of patients with multiple sclerosis (MS) employing virtual visual feedback and auditory feedback are reported in Chapters 18 and 19, respectively. The measured advantage of the checkerboard tile geometry over the transverse line geometry, as recorded in MS patients, is reported in Chapter 20. Test results recorders in children with cerebral palsy (CP) employing visual and auditory feedback are reported in Chapter 21. Finally, test results recorded in patients with Senile Gait (SG) and patients with previous stroke(s) (PS) are reported in Chapter 22. The concluding epilogue is followed by an Appendix laying out proofs of new theorems presented throughout the book chapters and by a list of references to previous publications.

Chapter 2

Some Mathematical Preliminaries

2.1 Probability spaces, probability densities and likelihood ratios

The purpose of this chapter is to present the basic underlying mathematical setup for the first two parts of this book (Chapters 3–14). It is based on fundamentals presented in an earlier work (Baram, 1976). Yet, while it seems essential to define at the outset the mathematical concepts involved, we aim at simplifying the definitions as much as possible. Along the same lines, well-known mathematical results by others are presented without proofs, while proofs of results derived by the author and collaborators are deferred to the closing Appendix of this book.

A probability space is a triple (Ω, U, P), where Ω is the sample space, which is the set of possible outcomes, U is the event space, an event being a set of outcomes in the sample space, and P is a probability function which assigns each event a probability, that is a number between 0 and 1. A measurable space is a couple (Ω, U), where Ω is a sample space and U is a collection of subsets of Ω (called σ-algebra).

An observation sequence (z_n), $n \geq 1$, is a stochastic (random) process on a probability space, (Ω, U, P), with values in a measurable space (R^ℓ, B^ℓ) called the observation space. We shall be interested in such a space, where R^ℓ is the ℓ-dimensional Euclidean space and B^ℓ is a σ-algebra in R^ℓ. The parameter space S on (Ω, U) is a set such that for each $s \in S$ there exists a probability measure P_s. Let $T \equiv (* \cup S)$, where, obviously, $* \in T$, but $*$ need not belong to the set S. We call $*$ the true parameter and P_* the *true measure*. For each $s \in T$ we denote by E_s expectation taken with respect to P_s. We use the notation a.e. (almost everywhere) to denote events of P_* measure one. Events of P_s measure one will be denoted a.e. P_s.

The conditional expectation of a random variable x on (Ω, U, P), given $A \in U$ is a U-measurable random variable denoted $E^A(x)$ such that

$$E\{E^A(x)\} = E(x) \tag{2.0}$$

For each $s \in S$ we shall denote by E_s^A the conditional expectation given A, taken with respect to P_s.

If μ and υ are measures defined on (Ω, U), then μ is said to be absolutely continuous with respect to υ if, for any set $A \in U$, $\upsilon(A) = 0$ implies $\mu(A) = 0$. μ is said to be singular with respect to υ if it is not absolutely continuous with respect to υ (this is not a standard definition. For a definition of mutually singular measures, see (Rudin, 1976, p. 121).

Let $(U_n) \equiv (U_n(z^n))$ be the increasing family of σ-subalgebras (subsets of σ-algebras) of U, generated by

$$z^n \equiv (z_1, \ldots, z_n) \tag{2.1}$$

For each $s \in T$ and for each $n \geq 0$, let $P_{s,n}$ denote the restriction of p_s to U_n. Suppose that for each $n \geq 0$ the measures $P_{s,n}$ are absolutely continuous with respect to some measure λ_n defined on (Ω, U_n). Then

$$\mathrm{f}_{s,n} \equiv \frac{\partial P_{s,n}}{\partial \lambda_n}; s \in T \tag{2.2}$$

are the Radon–Nikodym derivatives (or densities (Nikodym, 1930)) between the respective measures. The likelihood ratio between the parameters $s, t \in T$ is defined as

$$h_{t,n}^s \equiv \frac{\partial P_{s,n}}{\partial P_{t,n}} = \frac{\mathrm{f}_{s,n}}{\mathrm{f}_{t,n}} \tag{2.3}$$

provided that $P_{s,n}$ is absolutely continuous with respect to $P_{t,n}$. When the time parameter n is included in the argument, we shall use the somewhat shorter notation

$$\mathrm{f}_s(a(n)) \equiv \mathrm{f}_{s,n}(a(n)); s \in T; a(n) \in U_n$$

in particular

$$\mathrm{f}_s(Z^n) = \mathrm{f}_{s,n}(Z^n); s \in T \tag{2.4}$$

$$\mathrm{h}_s^t(Z^n) \equiv \mathrm{h}_{s,n}^t(Z^n); s, t \in T \tag{2.5}$$

For any $c \in U_n$ and $b \in U_n$ such that $\mathrm{f}_{s,n}(b) \neq 0$ for all $s \in S$, the conditional densities of c given b are

$$\mathrm{f}_{s,n}(c|b) \equiv \frac{\mathrm{f}_{s,n}(c,b)}{\mathrm{f}_{s,n}(b)}$$

In particular

$$\mathrm{f}_s(z_n|Z^{n-1}) = \frac{\mathrm{f}_s(Z^n)}{\mathrm{f}_s(Z^{n-1})}; \quad s \in T \tag{2.6}$$

The conditional likelihood ratios are then defined as

$$\mathrm{h}_t^s\left(z_n|Z^{n-1}\right) \equiv \frac{\mathrm{f}_s(z_n|Z^{n-1})}{\mathrm{f}_t(z_n|Z^{n-1})}; s, t \in T \tag{2.7}$$

for any Z^n such that $\mathrm{f}_t(z_n|Z^{n-1}) \neq 0$ for all t ∈ T.

The following condition will be assumed throughout the book:

(c2.1) For all $s \in S$, the probability measures $P_{s,n}$ are mutually absolutely continuous.

2.2 Bayesian probability densities

As before, consider a measurable space (Ω, U), where Ω is some sample space and U is a σ-algebra of subsets of Ω. Also consider a measurable space (S, U^s), where S is the parameter space and U^s is a σ-algebra of subsets of S. Let (Ω^b, U^b) be a measurable space, where

$$\Omega^b \equiv \Omega \times S$$

and

$$U^b \equiv U \times U^s$$

are the Cartesian products of the respective sample spaces and σ-algebras. Let P^b be a measure on (Ω^b, U^b). We denote by E^b expectation and by E^{b^A}; $A \in U$ conditional expectation given A, taken with respect to P^b. We call the restriction P_0^b of P^b to (S, U^s) the *a priori* probability measure on (S, U^s). Suppose that P_0^b is absolutely continuous with respect to some measure υ_0 on (S, U^s), then the density

$$\mathrm{f}_0^b \equiv \frac{\partial P_0^b}{\partial \upsilon_0} \tag{2.8}$$

is well defined. In particular, we call

$$\mathrm{f}_0^b(s); s \in S \tag{2.9}$$

the *a priori* probability density on S with respect to the measure υ_0.

Let (z_n) be a stochastic process on the probability space (Ω^b, U^b, P^b) with values in a measurable space (R^ℓ, B^ℓ), and let $(U_n^b) \equiv (U_n^b(Z^n))$ be the increasing sequence of σ-algebras of U^b, generated by $Z^n \equiv (z_1, \ldots, z_n)$. Let $P_n^b; n \geq 1$ be the restriction of P^b to U_n^b and for each $n \geq 1$ let P_n^b be absolutely continuous with respect to the measure υ_n defined on (Ω^b, U^b). Then the density

$$\mathrm{f}_n^b \equiv \frac{\partial P_n^b}{\partial \upsilon_n} \tag{2.10}$$

is well defined. We shall be particularly interested in the *a posteriori* probability density of s, given z^n

$$\mathrm{f}^b(s|Z^n) \equiv \mathrm{f}_n^b(s|Z^n) \equiv \frac{\mathrm{f}_n^b(s, Z^n)}{\mathrm{f}_n^b(Z^n)}; n \geq 1 \tag{2.11}$$

assuming $\mathrm{f}_n^b(Z^n) \neq 0$.

Let the parameter set S be finite, i.e.,

$$S = \{s_j; j \in K \equiv (0, \ldots, p)\} \tag{2.12}$$

For each $j \in K$ let

$$1_{s_j} \equiv \begin{cases} 1 & s = s_j \\ 0 & s \neq s_j \end{cases}$$

Then

$$\upsilon_0 \equiv \sum_{j=0}^{p} 1_{s_j}(s)$$

is a measure ("the counting measure") on (S, U^s). Let λ be a measure on (Ω, U), then $\upsilon_n = \upsilon_0 \cdot \lambda; n \geq 1$ is the product measure on (Ω^b, U^b).

Suppose that P_n^b is absolutely continuous with respect to υ_n (i.e., the entire measure P_n^b is concentrated on the set $\Omega \times \{s_i; i \in K\}$) for all $n \geq 0$, then we have

$$\mathrm{f}_0^b(s) = \sum_{i=0}^{p} \mathrm{f}_0^b(s_i) 1_{s_j}(s) \tag{2.13}$$

and

$$\mathrm{f}_0^b(s, Z^n) = \sum_{i=0}^{p} \mathrm{f}_n^b(s_i, Z^n) 1_{s_j}(s) \tag{2.14}$$

Hence

$$\mathrm{f}_n^b(Z^n) = \int_S \mathrm{f}_n^b(s, Z^n) d\upsilon_0 = \sum_{i=0}^{p} \mathrm{f}_n^b(s_i, Z^n) = \sum_{i=0}^{p} \mathrm{f}^b(Z^n | s_i) \tag{2.15}$$

where we have applied Bayes' rule (e.g., (Joyce, 2003))

$$\mathrm{f}_n^b(s_i, Z^n) = \mathrm{f}_0^b(s_i) \mathrm{f}_n^b(Z^n | s_i) \tag{2.16}$$

Substituting Eqs. (2.15) and (2.16) into Eq. (2.11) yields for each $j \in K$

$$\mathrm{f}^b(s_j | Z^n) = \frac{\mathrm{f}_0^b(s_j) \mathrm{f}_n^b(Z^n | s_j)}{\sum\limits_{i=0}^{p} \mathrm{f}_0^b(s_i) \mathrm{f}_n^b(Z^n | s_i)} \tag{2.17}$$

Note that

$$\mathrm{f}_n^b(Z^n | s_j) = \mathrm{f}_{s_j}(Z^n) \tag{2.18}$$

where the right-hand side is defined by Eq. (2.4). Thus, for the finite parameter set S

$$\mathrm{f}^b(s_j | Z^n) = \frac{\mathrm{f}_0^b(s_j) \mathrm{f}_{s_j}(Z^n)}{\sum\limits_{i=0}^{p} \mathrm{f}_0^b(s_i) \mathrm{f}_{s_i}(Z^n)} \tag{2.19}$$

2.3 Parameter estimates and stochastic convergence

An estimate $\hat{s}_n$ on S is a U_n-measurable mapping from Ω onto S. A maximum likelihood (ML) estimate on S is an estimate $\hat{s}_n \in S$

such that

$$\{f_s(Z^n); s \in S\} \leq f_{\hat{s}_n}(Z^n)$$

A maximum *a posteriori* probability (MAP) estimate on S is an estimate $\hat{s}_n \in S$ such that

$$\{f^b(s|Z^n); s \in S\} \leq f^b(\hat{s}_n|Z^n)$$

Let S be linear. Then the least-squares (LS) estimate on S is an estimate $\hat{s}_n \in S$ such that

$$E^b\{(s'_n - *)^T(s'_n - *)\} \geq E^b\{(\hat{s}_n - *)^T(s - *)\}$$

for any estimate s'_n on S, where $(s'_n - *)^T$ denotes $(s'_n - *)$ transposed and $*$ denotes the true parameter, assumed to have the same dimension as s.

Let the true parameter be assumed to belong to a finite set $\{s_j \in R^m; j \in K\}$. Then the LS estimate on R^m at instant n is the conditional expectation

$$E^{b^{U_n}} s = \int_S s f^b(s|Z^n) dv_0 = \sum_{j=0}^{p} s_j f^b(s|Z^n) \tag{2.20}$$

A stochastic sequence (x_n) on (Ω, U, P) is said to converge almost everywhere (a.e.) to a random variable x on (Ω, U, P) if

$$\lim_{n\to\infty} x_n = x \text{ a.e.}$$

A stochastic sequence (x_n) on (Ω, U, P) is said to converge in the mean (or in L_1) to a random variable x on (Ω, U, P) if

$$\lim_{n\to\infty} E|x_n - x| = 0$$

A vector-valued stochastic sequence (x_n) on (Ω, U, P) is said to converge in the mean-square (in m.s. or in L_2) to a random vector x on (Ω, U, P) if

$$\lim_{n\to\infty} E|x_n - x|^2 = 0$$

where $|x| \equiv (x^T x)^{1/2}$

A sequence of parameter estimates $(\hat{s}_n)$ is said to be consistent a.e. or in the m.s. if it converges a.e. or in m.s. to the true parameter.

We now present without proofs three well known results from probability theory, which are used in this book. Here, and in the rest of the book, the end of a theorem, a corollary or a lemma is denoted Δ, while the end of a proof is denoted $\square$.

Theorem 2.1 (Jensem's inequality, e.g., (Jensen, 1906; Bauer, 1972, p. 322)). *Let x be a real integrable random variable on a probability space (Ω, U, P) with values in R^1, and let $g(x)$ be a convex integrable function on R^1, then*

$$g(Ex) \leq Eg(x)$$

Theorem 2.2 (Fatou's Lemma, e.g., (Bauer, 1972, p. 71)). *Let (x_n) be an integrable stochastic sequence on (Ω, U, P) such that $x_n \geq 0$ a.e. for all n, then*

$$E \lim_{n\to\infty} \inf x_n \leq \lim_{n\to\infty} \inf Ex_n \qquad \Delta$$

Theorem 2.3 (Lebesgue's dominated convergence theorem, e.g., (Chung, 1974, p. 421)). *Let (x_n) be an integrable stochastic sequence on (Ω, U, P). Then if*

$$\lim_{n\to\infty} x_n = x \text{ a.e.}$$

where x is an integrable random variable on (Ω, U, P) and if there exists some integrable random variable y on (Ω, U, P) such that

$$Ey < \infty$$

and

$$|x_n| \leq y \text{ a.e. for all } n$$

then

$$\lim_{n\to\infty} Ex_n = Ex \qquad \Delta$$

2.4 Martingales and martingale difference sequences

A martingale is a sequence of random variables (i.e., a stochastic process) for which, at a particular time, the conditional expectation of the next value in the sequence is equal to the present value, regardless of all prior values. Let (Ω, U, P) be a probability space and let (U_n) be an increasing family of σ-subalgebras in U. A U_n-measurable stochastic sequence (x_n) on (Ω, U, P) is called a U_n-martingale if for each n

$$\begin{aligned} &\text{(a)}\;\; E|x_n| < \infty \\ &\text{(b)}\;\; E^{U^{n-1}} x_n = x_{n-1}\,\text{a.e.} \end{aligned}$$

If the equality in (b) is replaced by $\leq$, then (x_n) is called U_n-supermartingale.

It can be shown (Doob, 1953, p. 93) that the likelihood ratio sequence $\left(\frac{\partial P_{s,n}}{\partial P_{*,n}}\right)$; $s \in S$, defined in Section 2.2 is U_n-martingales according to the measure P_*. Hence

$$\begin{aligned} & E_*^{U^{n-1}} \frac{\partial P_{n,n}/\partial P_{s,n-1}}{\partial P_{*,n}/\partial P_{*,n-1}} \\ & \quad = \frac{\partial P_{*,n-1}}{\partial P_{s,n-1}} E_*^{U^{n-1}} \frac{\partial P_{s,n}}{\partial P_{*,n}} = \frac{\partial P_{*,n-1}}{\partial P_{s,n-1}} \frac{\partial P_{s,n-1}}{\partial P_{*,n-1}} = 1 \end{aligned}$$

Consequently, we have by Eqs. (2.7), (2.6) and (2.2)

$$E_*^{U_{n-1}} \mathrm{h}_*^s(z_n|Z^{n-1}) = E_*^{U_{n-1}} \frac{\mathrm{f}_s(z_n|Z^{n-1})}{\mathrm{f}_*(z_n|Z^{n-1})} = 1 \,\text{for each}\, s \in S \qquad (2.21)$$

Theorem 2.4 (The martingale convergence theorem, e.g. (Chung, 1974, p. 334; Bauer, 1972, pp. 341–343)). *Let (x_n) be a U_n-martingale on (Ω, U, P) and let*

$$\sup_{n \geq 0} E x_n^+ < \infty$$

where

$$x_n^+ = \sup(x_n, 0)$$

Then (x_n) converges a.e. to a finite limit. Δ

Let (Ω, U, P) be a probability space and let (U_n) be an increasing family of σ-sub-martingales in U_*. A U_*-measurable stochastic sequence on (Ω, U, P) is called a U_n-martingale difference sequence if it is integrable and if

$$E^{U_{n-1}} x_n = 0 \text{ a.e.}$$

Let Y_n be a stochastic sequence on (Ω, U, P) and let U_n be a sequence of σ-sub-martingales of U, generated by $(y_1, \ldots, y_n)$. Then, clearly $(y_n - E^{U_{n-1}} y_n)$ is a U_n-martingale difference sequence. Also note that if (x_n) is a U_n-martingale difference sequence, then

$$x_n \equiv \sum_{m=1}^{n} x_m$$

is a martingale. Indeed

$$E^{U_{n-1}} x_n = \sum_{m=1}^{n} E^{U_{n-1}} x_m = \sum_{m=1}^{n-1} x_m + E^{U_{n-1}} x_n = \sum_{m=1}^{n-1} x_m = x_{n-1}$$

Δ

2.5 Stationarity and ergodicity

The purpose of this section is to provide definitions, which will be useful in subsequent chapters. It is not intended to elaborate on the definitions involved. For a thorough presentation of ergodic theory, the reader is referred to, e.g., (Doob, 1953; Halmos, 1956; Chacon and Ornstein, 1960).

Consider a probability space (Ω, U, P). A transformation T from Ω to U is said to be measure preserving if

$$P(T^{-1}A) = P(A)$$

for all $A \in U$.

Given a measure preserving transformation T, a U-measurable event A is said to be invariant if

$$T^{-1}A = A(x_n) \text{ on } (\Omega, U, P)$$

Let (x_n) be a stochastic sequence on (Ω, U, P) with values in (R^ℓ, B^ℓ), where R^ℓ is the ℓ-dimensional Euclidean space and B^ℓ is the σ-algebra of Borel (countable) sets of R^ℓ. Let B^ℓ_∞ be the σ-algebra of Borel sets of R^ℓ_∞, where $R^\ell_\infty \equiv R^\ell \times R^\ell \times \ldots$. Then (x_n) is said to be stationary if for each $k \geq 1$

$$P[(x_1, \ldots, x_n) \in C] = P[(x_{k+1}, x_{k+2}, \ldots) \in C]$$

for every $C \in B^{\ell}_{\infty}$.

A stationary sequence is said to be ergodic if every invariant event in U has probability zero or one. It can be shown, e.g., (Stout, 1974, p. 168), that (x_n) is generated by a measure preserving transformation T (the shift operator), i.e.,

$$x_n(\omega) = x_{n-1}(T\omega) \tag{2.22}$$

Let (x_n) be a vector valued stochastic sequence from (Ω, U, P) into (R^{ℓ}, B^{ℓ}), such that the probability density with respect to the Lebesque measure on (R^{ℓ}, B^{ℓ}) of (x_n) is Gaussian on R^{ℓ}, with

$$Ex_n = m_x, \text{ constant for all } n$$

and

$$E\{(x_n - m_x)(x_{n+k} - m_x)^T\} \text{ depends only on } k$$

Then (x_n) is a stationary Gaussian sequence.

Proposition 2.1 (Doob, 1953, p. 494). *A zero-mean stationary Gaussian process (x_n) is ergodic if and only if*

$$\lim_{n \to \infty} \frac{1}{n+1} \sum_{k=0}^{n} |R(k)|^2 = 0$$

where

$$R(k) \equiv E\{x_n x_{n+k}^T\}$$

and where $|R(k)|$ denotes the determinant of $R(k)$. The notion of ergodicity is encapsulated as follows.

Theorem 2.5 (The ergodic theorem, e.g., (Doob, 1953, p. 464; Halmos, 1956, p. 22; Weiner, 1949, p. 16)). *Let (x_n) be an ergodic sequence on (Ω, U, P) and $\mathrm{f}(x_n)$ be a U-measurable function, such that $E|\mathrm{f}(x_0)|$ is finite, then*

$$\lim_{n \to \infty} \frac{1}{n+1} \sum_{j=0}^{n} \mathrm{f}(x_j) = E\,\mathrm{f}\,(x_0). \qquad \Delta$$

The following theorem (a version of the central limit theorem of probability theory) is highly useful in the characterization of ergodic martingale difference sequences.

Theorem 2.6 (Billingesley, 1961). *Let* (x_n) *be an ergodic stochastic process on* (Ω, U, P)*, such that* $E\{x_1^2\}$ *is finite and* $E^{U_{n-1}}\{x_n\} = 0$ a.e., *i.e.,* (x_n) *is an ergodic martingale difference sequence. Then the distribution of* $n^{-1/2}\sum_{k=1}^{n} x_k$ *approaches the Gaussian distribution with mean zero and variance* $E\{x_1^2\}$. Δ

2.6 Metric spaces and stochastic metrics

Consider a set S and a real-valued function e on $S \times S$, which satisfies

$$(i)\quad e(s;s) = 0 \text{ for any s} \in \text{S}$$

$$(ii)\quad e(s;t) = e(t;s) \text{ for any } s,t \in S$$

$$(iii)\quad e(s;t) \leq e(s;r) + e(r;t) \text{ for any } s,t,r \in S$$

Then e is called a pseudo-metric on S. If, in addition to $(i), (ii)$ and (iii), e satisfies

$$(iv)\quad e(s;t) = 0;\ s,t \in S \text{ implies } s = t$$

then e is called a metric on S. The pair (S, e) is called a metric space (Rendic, 1906, pp. 1–74; Choundhary, 1983).

Now consider a probability space (Ω, U, P) and an increasing family of σ-subalgebras of U. Let (e_n) be a U_n-measurable sequence of functions on $S \times S$ such that each e_n satisfies $(i) - (iii)$ above. Then we shall call (e_n) a stochastic pseudo-metric sequence on S. If each e_n satisfies $(i) - (iv)$ above, we shall call (e_n) a stochastic metric sequence on S.

2.7 Information

In most of this book, probabilities serve as measures of information, which are ultimately associated with sequences. While probabilities may be viewed as representing a level of uncertainty, information represents a level of certainty. We start with the notion of information in a single observation. We do not generally make the assumption, common in the derivation information measures, that a true parameter is included in a given parameter set, or, equivalently, that the true measure is included in a known set of measures. Here, the mean and the conditional mean values of the discriminating information in a single observation are first shown to possess properties that will prove useful in the following chapters. In particular, their absolute values are shown to be metrics, or distance measures, on parameter spaces. This provides a meaningful measure of the relative closeness of parameters to the true parameter, which may not be included in the given parameter set.

Let S be a parameter space and $T = (* \cup S)$, where $*$ is the true parameter. If for some pair of parameters $s, t \in T$

$$\mathrm{f}_s(Z^n) > \mathrm{f}_t(Z^n)$$

or, equivalently,

$$\log\, \mathrm{f}_s(Z^n) > \log \mathrm{f}_t(Z^n)$$

we say that the parameter s is favored over the parameter t by the observations Z^n. Then $\log \mathrm{f}_s(Z^n)$ may be regarded as a measure of the information in Z^n for selecting a parameter from the set T. The difference

$$\log \mathrm{f}_s(Z^n) - \log \mathrm{f}_t(Z^n) = \log \mathrm{h}_t^s(Z^n) \tag{2.23}$$

is then a measure of the information in Z^n for selecting between s and t. If Eq. (2.23) is positive, then s is favored, and if it is negative, then t is favored. The difference

$$\log \mathrm{h}_t^s(Z^n) - \log \log \mathrm{h}_t^s(Z^{n-1}) = \log \mathrm{h}_t^s(z_n|Z^{n-1}) \tag{2.24}$$

is then a measure of the difference between the information favoring s over t at instance $n-1$. It can be regarded as a measure of the information favoring s over t in the observation z_n. We define

$$I_n(s;t) \equiv E_*^{U_{n-1}} \log \mathrm{h}_t^s\left(z_n|Z^{n-1}\right) \tag{2.25}$$

as the conditional mean information in z_n favoring s against t and

$$\bar{I}_n(s;t) \equiv E_* \log \mathrm{h}_t^s(z_n|Z^{n-1}) \tag{2.26}$$

as the mean information in z_n favoring s against t. A more general form of Eq. (2.26) would be $\tilde{I}_n(s;t) \equiv E_*^{A_{n-1}} \log \mathrm{h}_t^s(z_n|Z^{n-1})$ for some sequence (A_n), such that $A_n \in U_n$.

However, for the purposes of this book, we use the information defined by Eq. (2.26).

2.8 Properties of information

We now show some properties of the information measures defined above that will prove to be useful in the following chapters. The proofs of Theorems 2.6–2.8 and corollaries 2.1, 2.2 below are provided in the Appendix of this book.

Theorem 2.7. *Let S be a parameter space. Then for any $s \in S$ and for each $n \geq 0$, we have*

$$I_n(*;s) \geq 0 \text{ a.s.} \tag{2.27}$$

and

$$\bar{I}_n(*;s) \geq 0$$

with equality if and only if $\mathrm{f}_s(z_n|Z^{n-1}) = \mathrm{f}_*(z_n|Z^{n-1})$ *a.e.* Δ

Theorem 2.8. *The sequence $(|\bar{I}_n(s;t)|); s,t \in S$ is a sequence of pseudo metrics on S. It is a sequence of metrics on S if and only if $\bar{I}_n(s;t) = 0$ implies $s = t$. The sequence $(|I_n(s;t)|); s,t \in S$ is a stochastic sequence of pseudo metrics on S. It is a sequence of metrics if and only if* $I_n(s;t) = 0$ implies $s = t$. Δ

Theorem 2.9. *For any $t, r \in S$ and each $n \geq 0$, the sequences $(|\bar{I}_n(*;t)|)$ and $(|I_n(*;t)|)$ satisfy the properties*

$$\begin{aligned} (i)&\ |\bar{I}_n(s;s)| = 0 \text{ for any } s \in S \\ (ii)&\ |\bar{I}_n(s;t)| = |\bar{I}_n(t;s)| \text{ for any } s,t \in S \\ (iii)&\ |\bar{I}_n(s;t)| \leq |\bar{I}_n(s;r)| + |\bar{I}_n(r;t)| \text{ for any } s,t,r \in S \end{aligned}$$

They satisfy

$$(iv)\ |\bar{I}_n(s;t)| = 0 \text{ for any } s,t \in S \text{ implies } s = t$$

if and only if $\mathrm{f}_t(z_n|Z^{n-1}) = \mathrm{f}_*(z_n|Z^{n-1})$ *a.e. implies* $t = *$. Δ

Corollary 2.1. *Suppose that* $r \in S$ *is the true parameter. Then for any* $t \in S$, $I_n(s;t)$ and $\overline{I}_n(s;t)$ *are maximized on* S at $s = r$. *This maximum is unique unless for some* $s \in S$, $f_s(z_n|Z^{n-1}) = f_s(z_n|Z^{n-1})$ *a.e.* △

Corollary 2.2. *Let* $s, t \in S$ *be any pair of parameters in the parameter space* S. *Then* s *is closer to the true parameter* $*$ *than* t *in the metric* $|I_n(s;t)|$ *if and only if* $I_n(s,t) > 0$ *a.e. and in the metric* $|\overline{I}_n(s;t)|$ *if and only if* $\overline{I}_n(s;t) > 0$. △

PART I

Sensory Feedback Control: Optimal State-Space Design and Analysis

Chapter 3

Minimal-Order State Estimation of Feedback Linear Systems

3.1 Introduction

The minimality of required resources would seem to be a natural preference in the evolutionary design of biological mechanisms. From a scientific viewpoint, the establishment of a mathematical theory supporting such evolutionary preference would appear to be particularly rewarding. State estimation seems to be a fundamental necessity underlying biological movement. Order reduction, and even minimality, of the state model (or state estimator) involved implies, on the one hand, stationarity of the model, and, on the other, nonexistence of a lower-order state estimator. Stationary linear feedback models have been adopted in the analysis of optogenetically driven neural activity, e.g., (Nataraj and van den Bogert, 2017). However, a mathematically reliable, effective biological movement control paradigm would appear to require a further look. Specifically, even a well-thought linear control model would need to be supported by facilities allowing identification of, and adaptation to, varying external and internal conditions. Employing linear feedback systems, control theory suggests that order reduction can be obtained when the observation noise is colored or when it is white with a nonsingular covariance matrix (Kalman, 1960b; Bryson and Johansen, 1965). Here, following an earlier observation (Baram and Shaked, 1986), we show that order reduction can be enforced by the internal structure of the system, even when the observation noise is white with a nonsingular covariance. Furthermore, assuming a nonsingular observation noise covariance, we obtain a necessary and sufficient condition for the minimality of the state estimator in terms of the internal system structure and parameters. These, and the following theoretical results presented in the next two chapters, are laid out for the benefit of the mathematically inclined reader.

3.2 Linear system model

Consider the system

$$x_{n+1} = Ax_n + Bw_n, \quad x_n \in R^p \tag{3.1a}$$

$$y_n = Cx_n + u_n, \quad y_n \in R^q \tag{3.1b}$$

with

$$E\{x_0\} = E\{w_n\} = E\{u_n\} = 0, \quad E\{w_n x_0^T\} = E\{u_n x_0^T\} = 0$$

with

$$E\{w_n w_k^T\} = Q\delta_{n,k}, \quad E\{u_n u_k\} = P\delta_{n,k}$$

where E denotes estimation operator with

$$Q > 0, P > 0 \quad \text{and} \quad \delta_{n,k} = \begin{cases} 1 & n = k \\ 0 & n \neq k \end{cases}$$

We further assume

$$E\{w_n u_k^T\} = 0$$

Let us denote by $Y_n^- = \{y_{n-k}, k \geq 0\}$ the space of past observations at time n and by $x_{n|k}$ the linear mean-square projection of x_n on Y_k^-. We assume that the system is asymptotically stable and that $E\{x_{n|n-1}x_{n|n-1}^T\}$ has a constant value Π and $E\{(x_n - x_{n|n-1})(x_n - x_{n|n-1})^T\}$ has a constant value Σ. It is well known that Π and Σ satisfy the equations

$$\Pi - A\Pi A^T = AKRK^T A^T \tag{3.2}$$

$$\Sigma - A\Sigma A^T + AKC\Sigma A^T = BQB^T \tag{3.3}$$

where

$$K = \Sigma C^T R^{-1}, \quad R = C\Sigma C^T + P$$

The state predictions are obtained from the stationary Kalman filter (Kalman, 1960)

$$x_{n+1|n} = Ax_{n|n-1} + AK\nu_n \tag{3.4}$$

where $v_n = y_n - Cx_{n|n-1}$.

Let us denote by T the matrix whose columns are the eigenvectors of Π corresponding to nonzero eigenvalues and define

$$z_n = T^T x_{n|n-1} \tag{3.5}$$

Then z_n has orthonormal components and dim z_n = rank Π. Furthermore, we have the inverse transformation

$$x_{n|n-1} = Tz_n \tag{3.6}$$

Substituting Eqs. (3.5) and (3.6) into Eq. (3.4), we get

$$z_{n+1} = T^T ATz_n + T^T AK\nu_n \tag{3.7}$$

with $\nu_n = y_n = CTz_n$. Estimation updates can be obtained as

$$x_{n|n} = Tz_n + K\nu_n \tag{3.8}$$

Since, clearly, there is no vector of dimension lower than that of z_n from which $x_{n|n-1}$ can be obtained by a linear transformation, it seems justified to call Eq. (3.7) a minimal state estimator of the system Eq. (3.1).

3.3 Filter order minimality

Recalling that to each Jordan block $J_i, i = 1, \ldots, t$, in the spectral decomposition of A there corresponds one eigenvalue α_i and one eigenrow v_i^T (we note that α_i is not necessarily different from α_j for $i, j = 1, \ldots, t$). We have the following result.

Theorem 3.1. *The filter Eq. (3.4) is of minimal order if and only if (A, B) is reachable, A is not singular and for any eigenrow v_i^T of A*

$$C(\alpha_i^{-1} I - A)^{-1} BQB^T v_i \neq 0 \tag{3.9}$$

Δ

Let us now examine the conditions of Theorem 3.1. The matrix A is known to be nonsingular if the system is obtained by discretizing a continuous-time linear system.

Suppose that the pair (A, B) is reachable. Then, by the Popove–Belevitch–Hautus test, e.g., (Hespanha, 2018), $B^T v_i \neq 0$ for all i. Suppose further that $r \geq q$, i.e., that there are no less outputs than inputs. Then, a sufficient condition for minimality of the filter is that the matrix $C(\alpha_i^{-1}I = A)^{-1}B$ is nonsingular. It follows that a sufficient condition for filter minimality in this case is that α_i^{-1} is not a zero of the system, i.e., that none of the system's zeros is a reciprocal of any of its poles. Reciprocal pole-zero pairs are known as all-pass elements as they represent a flat (unity) spectral density. In the single-input, single-output case, we immediately obtain that when the pair (A, B) is reachable, a necessary and sufficient condition for filter minimality is that the system contains no all-pass pairs. Condition Eq. 3.9 may then be regarded as a generalization of the no-all-pass condition for the multivariable case.

Finally, we note that the concept of filter order reduction, in the sense of this chapter, is different from that of degeneracy of the Ricatti equation, which has received considerable attention in the literature, e.g., (Gevers and Kailath, 1973; Silverman, 1976). The minimal filter order is equal to the rank of the matrix Π and not to that of the matrix Σ. Consequently, the filter order may not be minimal even when Σ is of full rank, as condition Eq. (3.9) need not be satisfied even if the system is reachable and observable. In the single-input, single-output case, when the system is reachable and observable and contains all-pass elements, Σ will be of full rank but Π will be rank-deficient, yielding a state estimator of order lower than that of the given system.

3.4 Discussion

We have investigated the relationships between a stochastic linear system in state-space form and its minimal-order state estimator. A necessary and sufficient condition for the minimality of the Kalman filter as a state estimator was obtained in terms of the internal system structure, which appears to be naturally manifested by biological constructs, such as vision and gait entrainment. Another context of state-space minimality, namely, error minimality, is addressed in the following chapter. Both concepts will be shown to be highly relevant in biological contexts, where neurological impairment is compensated by technological support. Specifically, the technological choice of closed-loop feedback control, which is clinically found to be highly superior to open-loop control, is hereby asserted on mathematical grounds. This seems to be a suitable meeting-place of mathematical theory and medical practice.

Chapter 4

Estimability and Regulability of Linear Systems

4.1 Introduction

The benefit of using observation or feedback signals in state estimation or regulation of dynamical systems is normally manifested by the reduction of certain cost functions with respect to their values when no such signals are used. In this chapter we introduce those properties of linear systems that yield mean-square error reduction in state estimation and quadratic cost reduction in state feedback control. These properties, which we call estimability and regulability (Baram and Kailath, 1988), are different from the properties of observability and controllability, which were introduced by Kalman (1960a; 1960b) and are widely recognized as key structural properties in linear estimation and control.

We shall say that a stochastic linear system is estimable if, in estimating its state from its output, the posterior error covariance matrix is strictly smaller than the prior state covariance matrix. This condition is equivalent to the condition that no direction in the state space at any time is orthogonal to all the past observations. It is also shown to be equivalent to the reachability of a certain pair of matrices related to the covariance function of the observation process. The notion of estimability is independent of that of observability in the sense that one does not imply the other. When a system is not estimable, the corresponding Kalman filter can be reduced to a lower order estimator of the state process. A linear state-space system driven by white noise is shown to be a minimal-order realization of its output process if and only if it is observable and estimable. These results apply to time-variant and time-invariant linear systems in discrete and continuous time. In the stationary case, the estimability condition leads directly to conditions for minimality of the Kalman filter in discrete (Baram and Shaked, 1984; 1986; 1988) and continuous (Baram and Shaked, 1985) time.

We shall say that a linear system is regulable if, for any nonzero initial condition, the quadratic cost in applying optimal control is strictly smaller than the cost when no control is applied. This means that when a system is not regulable, the feedback signal may be eliminated for some nonzero initial condition without increasing the cost. Regulability is shown to be equivalent to a certain reachability condition, which was shown to be equivalent to the irreducibility of the optimal regulator (Bernhard and Cohen, 1974). We close by showing that, as might be expected, regulability and estimability are dual properties.

4.2 Estimability

A stochastic linear feedback control system is said to be estimable if, in estimating its state from its output, the posterior error covariance matrix is strictly smaller than the prior state covariance matrix.

Consider the system

$$\begin{aligned} x_{k+1} &= F_k x_k + G_k w_k \quad & x_k \in R^n \\ y_k &= H_k x_k + v_k & k = 0, 1, 2, \cdots \end{aligned} \tag{4.1}$$

where $E\{x_0\} = 0, E\{x_0 x_0^T\} = \Pi_0, \{w_k\}$ and $\{v_k\}$ are zero-mean sequences with $E\{x_0 w_k^T\} = 0$, $E\{w_k v_j^T\} = 0$, $E\{w_k w_j^T\} = Q_k \delta_{k,j}$, $E\{v_k v_j^T\} = R_k \delta_{k,j}$. Also, $E\{\cdot\}$ denotes expectation, $(\cdot)^T$ denotes transpose, and $\delta_{k,j}$ denotes the Kronecker delta. Let $\hat{x}_k$ denote the linear least-mean-square estimate of x_k given $\{y_k, y_{k-1}, \ldots\}$, and let

$$\Pi_k = E\{x_k x_k^T\}$$

and

$$P_k = E\{(x_k - \hat{x}_k)(x_k - \hat{x}_k)^T\}$$

The elements of x_k and y_k may represent, in technological settings, such entities as measures of aircraft control or movement parameters (e.g., pitch rate, elevator angle, flying speed and wind gust, etc., as in, e.g., (Baram and Eidelman, 1985)) or, as in the biological sensorimotor settings considered in subsequent chapters, such measures as line of sight angle in descending bird or walking speed and stride length in neurologically impaired patients).

Definition 4.1. The system Eq. (4.1) is estimable if

$$P_k < \Pi_k \quad \text{for all k} \geq \text{n}$$

i.e., if and only if for any $g \in R^n, g \neq 0$

$$g^T P_k g < g^T \Pi_k g \quad \text{for all k} \geq \text{n} \tag{4.2}$$

or, in words, the posterior mean-square error in estimating any linear combination of the state variables is strictly smaller than the prior mean-square error. As will be shown, the requirement $k \geq n$ is needed since, for $k < n$, the condition will immediately fail for, say, a scalar $\{y_k\}$ sequence.

The orthogonality property of the least-mean-square estimate implies that Eq. (4.2) will fail if and only if $g^T x_k$ is orthogonal to (uncorrelated with) the sequence $\{y_j, j \leq k\}$ for $k \geq n$. Conversely, Eq. (4.2) is equivalent to the condition

$$g^T E\{x_k y_j^T\} \neq 0 \quad \text{for some } j \leq k \text{ and all } k \geq n \tag{4.3}$$

which means that no direction in the state space for $k \geq n$ is orthogonal to all the past observations.

Noting that

$$x_k = \Phi_{k,j} x_j + \sum_{i=0}^{k-j-1} \Phi_{k,j+i+1} G_{j+i} w_{j+i}$$

where $\Phi_{k,j} = F_{k-1} F_{k-2} \cdots F_j, j \leq k-1, \Phi_{k,k} = I$ denotes the system's state transition matrix, the estimability condition Eq. (4.3) can be written as

$$y^T\Phi_{k,j}N_j \neq 0 \quad \text{for some } j \leq k \text{ and all } k \geq n \tag{4.4}$$

where

$$N_j = \Pi_j H_j^T \tag{4.5}$$

Condition Eq. (4.4) holds if and only if the matrix sequence $\{\Phi_{k,j}N_j, j \leq k\}$ has column rank n for all $k \geq n$, or, equivalently, the estimability Grammian

$$W_k = \sum_{j=k}^{0} \Phi_{k,j}N_jN_j^T\Phi_{k,j}^T \tag{4.6}$$

has full rank for all $k \geq n$. It can now be seen that we require $k \geq n$ in the definition of estimability since for a scalar $\{y_k\}$ sequence, the matrix sequence $\{\Phi_{k,j}N_j, j \leq k\}$ cannot have column rank n for any $k < n$ and the condition will immediately fail. We note that the triple $\{H_k, \Phi_{k,j}, N_j, j \leq k\}$ defines the covariance function of $\{y_k\}$

$$\begin{aligned} R_y\{k,j\} = E\{y_ky_j^T\} &= (R_k + H_k\Pi_kH_k^T)\delta_{k,j} \\ &\quad + H_k\Phi_{k,j}N_j1(k-j) + N_k^T\Phi_{k,j}^TH_j^T1(k-j) \end{aligned}$$

where $1(k-j) = 1$ for $j < k, 0$ otherwise, and Π_k satisfies the Lyapunov equation

$$\Pi_{k+1} = F_k\Pi_kF_k^T + G_kQG_k^T \tag{4.7}$$

We have, then, the following result

Theorem 4.1. *The system Eq.* (4.1) *is estimable if and only if*

$$\text{rank } W_k = n \quad \text{for all } k \geq n \tag{4.8}$$

Remarks

1) As can be seen from the form of W_k, estimability is equivalent to reachability of the pair $\{F_k, N_k\}$.
2) When the system is not estimable, the state estimate can be reduced, without loss of information, to a lower dimensional vector, which, in turn, results in the reduction of the Kalman

filter. Such reduction can be accomplished by the transformation $z_k = T_k \hat{x}_k$, where T_k is the matrix whose columns are the eigenvectors of $E\{\hat{x}_k \hat{x}_k^T\} = \Pi_k - P_k$ corresponding to its nonzero eigenvalues.

3) The concept of estimability may be regarded as an extension to the stochastic case of the concept of observability, which, in the deterministic case, means that the state can be completely deduced from the past observations and inputs. This notion of stochastic observability is different from the one defined by Chen (1985) as convergence of the error covariance matrix in estimating the initial state from a finite number of observations, when the prior covariance matrix is αI and α tends to infinity. While Chen's definition reduces to that of observability when the system is deterministic, the estimability notion is independent of that of observability, in the sense that one does not imply the other (see the example below).

4) Motivated by Eq. (4.3), let us examine the condition that no direction in the state space is orthogonal to all the past inputs (rather than outputs, as in Eq. (4.3)), i.e., that for any $g \in R^n$, $g \neq 0$

$$g^T E\{x_k w_j^T\} \neq 0 \quad \text{for some } j \leq k-1 \text{ and all } k \geq n$$

which, noting that

$$x_k = \Phi_{k,0} x_0 + \sum_{j=0}^{k-1} \Phi_{k,j+1} G_j w_j$$

becomes

$$g^T \Phi_{k,j+1} G_j Q_j = 0 \quad \text{for some } j \leq k-1 \text{ for all } k \geq n \tag{4.9}$$

for any $g \in R^n, g \neq 0$. But the latter condition holds if and only if the matrix sequence $\{\Phi_{k,j} G_j Q_j, j \leq k-1\}$ has full column rank, which is the condition for reachability of the pair $(F_j, G_j Q_j)$. This condition, which has been termed disturbability (Bryson, 1978), is identical to the deterministic reachability condition when the output covariance Q_j is nonsingular. This stands in contrast to the fact that estimability and observability are unrelated concepts.

We also note that disturbability is not dual to estimability in the common system-theoretic sense.

4.2.1 *The stationary case*

Some useful equivalent forms of Theorem 4.1 can be obtained in the special case where the parameters $\{F, G, H, Q, R\}$ of the system 4.1 are time invariant. When the eigenvalues of F are strictly inside the unit circle, the matrix $\Pi_k = E\{x_k x_k^T\}$ has a steady-state value Π, which is the unique nonnegative solution of the equation (see, e.g., (Anderson and Moore, 1979, p. 70))

$$\Pi = F\Pi F^T + GQG \tag{4.10}$$

If $\Pi_0 = \Pi$, then Π_k has a constant value Π for all $k \geq 0$ and the processes x_k and y_k are stationary. The matrix $N_k = \Pi_k H_k^T$ then has a constant value $N = \Pi H^T$. The stationary system is estimable if the condition

$$P_k < \Pi \quad \text{for all } k \geq n$$

is satisfied. The following result is a direct consequence of Theorem 4.1.

Corollary 4.1. *The stationary system* $\{F, G, H, Q, R, \Pi_0 = \Pi\}$ *is estimable if and only if the pair* (F, N) *is reachable, i.e., rank* $[N : FN : \ldots : F^{n-1}N] = n$.

The Popov–Belevitch–Hautus test (PBH), e.g., (Hespanha, 2018), can be used to obtain an equivalent characterization of estimability. Let us denote by f_j^T a left eigenvector of F corresponding to an eigenvalue λ_i, then $f_i^T F = \lambda_i f_i^T$. Multiplying Eq. (4.10) on the right by f_i, we obtain

$$\Pi f_i = (I - \lambda_i F)^{-1} GQG^T f_i$$

yielding

$$N^T f_i = H(I - \lambda_i F)^{-1} GQG^T f_i \tag{4.11}$$

By the PBH test, the pair (F, N) is reachable if and only if $N^T f_i \neq 0$ for any f_i corresponding to each λ_i. Noting that $\lambda_i \neq 0$, we obtain, by Eq. (4.11) and Corollary 4.1, the following result, which relates the estimability property to the system's eigenstructure.

Corollary 4.2. *The stationary system* $\{F, G, H, Q, R, \Pi_0 = \Pi\}$ *is estimable if and only if for any left eigenvector* f_i^T *corresponding to each eigenvalue* λ_i *of* F,

$$H^T(\lambda_i^{-1}I - F)^{-1}GQG^T f_i \neq 0 \tag{4.12}$$

Condition Eq. (4.12) is the same as the condition presented earlier (Baram and Shaked, 1985) for minimality of the stationary Kalman filter associated with the system, in the sense that there is no linear estimator of its state, having order smaller than n. The above derivation is considerably simpler than that previous one.

The definition of estimability and the conditions obtained above are readily extendable to the continuous-time case, where the estimability Grammian is defined in complete analogy to the discrete-time case, and the term $(\lambda_i^{-1}I - F)$ is replaced by the term $(-\lambda_i I - F)$.

Finally, we give a simple example to illustrate that estimability does not imply and is not implied by observability.

Example: Consider a stationary system of the form Eq. (4.1) with

$$A = \begin{bmatrix} -\frac{1}{2} & 0 \\ 0 & -\frac{1}{4} \end{bmatrix}, \quad B = \begin{bmatrix} -6 \\ 7 \end{bmatrix}, \quad C = [1 \quad 1], \quad Q = 1$$

It can be readily verified that the system is observable and reachable, yet, for $\lambda_i = -1/2$, we have

$$C(\lambda_i^{-1}I - A)^{-1}B = [1\ 1] \begin{bmatrix} -\frac{2}{3} & 0 \\ 0 & -\frac{4}{7} \end{bmatrix} \begin{bmatrix} -6 \\ 7 \end{bmatrix} = 0$$

hence, condition Eq. (4.12) is violated, implying that the system is not estimable. On the other hand, taking

$$C = [1 \quad 0]$$

and noting that

$$f_1 = \begin{bmatrix} 1 \\ 0 \end{bmatrix} \quad \text{and} \quad f_2 = \begin{bmatrix} 0 \\ 1 \end{bmatrix}$$

we obtain

$$C(\lambda_i^{-1} I - A)^{-1} BQB^T f_1 = 24 \neq 0$$

and

$$C(\lambda_i^{-1} I - A)^{-1} BQB^T f_2 = 12 \neq 0$$

implying that the system, which is unobservable, is estimable.

4.2.2 *Minimal order realization of the observation process*

It is known (see, e.g., (Faurre, 1976)) that the minimal order of any realization of a stationary process $\{y_k\}$, possessing a finite dimensional linear representation, is the dimension of the Hilbert space Y_k generated by $\{\hat{y}_{k+j}, j \geq 0\}$, where $\hat{y}_{k+j}$ denotes the least-squares estimate of y_{k+j} given $\{y_{k-1}, y_{k-2}, \ldots, y_0\}$. To see that this is also true when $\{y_k\}$ is nonstationary, let z_k denote a base of Y_k, i.e., a vector of minimal dimension from which any element of Y_k can be obtained by a linear transformation. Since $z_{k+1} \in Y_{k+1} \subset Y_k \otimes \nu_k$, where $\nu_k = y_k - \hat{y}_k$ and $\otimes$ denotes the Cartesian product, and, since $\nu_k \perp Y_k$, there exist matrices A_k, B_k and C_k, such that $z_{k+1} = A_k z_k + B_k \nu_k$ and $y_k = C_k z_k + \nu_k$. By definition, $\dim z_k = \dim Y_k$. Suppose that there exists a realization of $\{y_k\}$, given by $q_{k+1} = \alpha_k q_k + \beta_k r_k, y_k = \gamma_k q_k + t_k$, where $\{r_k\}$ is a white noise process uncorrelated with q_0, and t_k is a white noise process uncorrelated with $\{r_k\}$, such that $\dim q_k < \dim Y_k$. Then we have $[\hat{y}_k, \hat{y}_{k+1} \ldots] = [\gamma_k, \gamma_{k+1}\theta_{k+j,k}, \ldots]\hat{q}_k$, where $\theta_{k+j,k}$ is the transition matrix corresponding to α_k and $\hat{q}_k$ is the least-square estimate of q_k, given $[y_{k-1}, y_{k-2}, \ldots]$. But since the dimension of Y_k is equal to the covariance rank of $[\hat{y}_k^T : \hat{y}_{k+1}^T : \ldots]^T$, we have $\dim Y_k \leq \dim q_k < \dim Y_k$, a contradiction. It follows that the minimal order of any realization of $\{y_k\}$ at time k is equal to the dimension of Y_k.

Since, by Eq. (4.1), we have $[\hat{y}_k, \hat{y}_{k+1}, \ldots] = [H_k, H_{k+1}\Phi_{k+1,k}, \ldots]\hat{x}_k$, it follows that $\dim Y_k = n$ if and only if the system Eq. (4.1) is observable and rank $E\{\hat{x}_k\hat{x}_k^T\} = n$. Since the latter, for $k \geq n$, is the condition for estimability, we have the following result.

Theorem 4.2. *The system Eq.* (4.1) *is a minimal-order realization of the process* $\{y_k\}$ *for* $k \geq n$ *if and only if it is observable and estimable.*

System estimability implies, by definition, that the observations contain information on all the state variables. It might be expected, then, that the minimality of any realization of the observation process would be strongly linked to the estimability property. The fact that estimability depends on the sequence $\{N_k\}$, which defines the covariance function of the output process, is a further indication of such a linkage. We first note that if we define the order of a state-space realization of a possibly nonstationary process as the dimension of its state process, then the order will be generally time-dependent. We shall say that a realization of such a process is minimal if its order is not greater than that of any other realization of the process at any time. Finally, we note that the early notion of estimability (Baram and Kailath, 1988) has been later linked to that of "complete stochastic observability" (Liu and Bitmead, 2011), which, suggesting that the term "complete" implies that no non-trivial function of the state is unobservable in turn, has been also linked to the notion of "conditional entropy" (Mohler and Hwang, 1988).

4.3 Regulability

The problems of linear mean-square estimation and linear quadratic regulation are known to be analogous. It is therefore natural to explore the analogy of estimability, considered in the previous section, in the regulation problem. We call this analog regulability. A stochastic linear feedback control system is said to be regulable if, for any nonzero initial condition, the quadratic cost in applying optimal control is strictly smaller than the cost when no control is applied. The linear quadratic regulator reduces the amount of work done by the control system. This means that, when a system is not regulable, the feedback signal may be eliminated for some nonzero initial condition without increasing the cost. The cost function is normally defined as a sum of the deviations of key measurements from their desired values.

Consider the system

$$x_{k+1} = F_k x_k + G_k u_k \tag{4.13}$$

where the input sequence $\{u_k\}$ is to be selected so as to minimize the quadratic cost function

$$J\{x_0, \{u_k\}, K\} = \sum_{k=0}^{K-1} x_{k+1}^T L_{k+1} K_{k+1} + u_k^T M_k u_k \tag{4.14}$$

where L_k is a nonnegative definite matrix and M_k is a positive definite matrix.

Definition 4.2. The system Eq. (4.13) is regulable if for any initial condition $x_0 \neq 0$

$$\min_{\{u_k\}} J[x_0, \{u_k\}, K] < J[x_0, \{u_k = 0\}, K] \quad \text{for all } K \geq n \tag{4.15}$$

Regulability means that for any initial condition, the optimal cost is strictly smaller than the cost when no input is used.

We next derive a Gramian rank test for regulability. Substituting

$$x_k = \Phi_{k,0} x_0 + \sum_{j=0}^{k-1} \Phi_{k,j+1} G_j u_j$$

in Eq. (4.14), we obtain

$$J\{w_0, \{u_k\}, K\}$$
$$= \sum_{k=0}^{K-1} \left\{ x_0^T \Phi_{k+1.0}^T L_{k+1} \Phi_{k+1,0} x_0 \right.$$
$$+ 2x_0^T \Phi_{k+1.0}^T L_{k+1} \sum_{j=0}^{k} \Phi_{k+1,j+1} G_j u_j$$

$$+\left(\sum_{j=0}^{k}\Phi_{k+1,j+1}G_j u_j\right)^T L_{k+1}\cdot\left(\sum_{j=0}^{k}\Phi_{k+1,j+1}G_j u_j\right)$$

$$\left. +u_k^T M_k u_k \right\}$$

Let

$$\mathrm{A}_{k,j} = L_{k+1}^{1/2}\Phi_{k+1,j+1}G_j, \quad b_k = L_{k+1}^{1/2}\Phi_{k+1,0}x_0$$

and

$$\mathrm{A} = \begin{bmatrix} \mathrm{A}_{0,0} & & & & \\ \mathrm{A}_{1,0} & \mathrm{A}_{1,1} & & & \\ \mathrm{A}_{2,0} & \mathrm{A}_{2,1} & \mathrm{A}_{2,2} & 0 & \\ & & & \ddots & \\ \vdots & & & & \\ \mathrm{A}_{K-1,0} & \mathrm{A}_{K-1,1} & \mathrm{A}_{K-2,2} & \cdots & \mathrm{A}_{K-1,K-1} \end{bmatrix} \quad b = \begin{bmatrix} b_0 \\ b_1 \\ \vdots \\ b_{k-1} \end{bmatrix}$$

$$M = \begin{bmatrix} M_1 & & \\ 0 & \ddots & 0 \\ & & M_K \end{bmatrix} \quad u = \begin{bmatrix} u_0 \\ u_1 \\ \vdots \\ u_{k-1} \end{bmatrix}$$

Then it can be verified that

$$\begin{aligned} \Delta J &= J[x_0, \{u_k\}, K] - J[x_0, \{u_k = 0\}, K] \\ &= u^T[\mathrm{A}^T\mathrm{A} + \mathrm{M}^T]u + 2b^T\mathrm{A}u \end{aligned} \tag{4.16}$$

We would like to find a necessary and sufficient condition for regulability, i.e., a condition under which $\min_{\{u_k\}} \Delta J < 0$. Obviously, we have $\min_{\{u_k\}} \Delta J \leq 0$, so we would like to find a condition under which

$$\min_{\{u_k\}} \Delta J \neq 0 \tag{4.17}$$

Differentiating Eq. (4.16) and equating to zero, we obtain

$$\frac{\partial \Delta J}{\partial u} = 2u^T(\mathrm{A}^T\mathrm{A} + \mathrm{M}) + 2b^T\mathrm{A} = 0$$

Since M has full rank, so does $\mathrm{A}^T\mathrm{A}+\mathrm{M}$. Hence, the optimal input is

$$u = -(\mathrm{A}^T\mathrm{A}+\mathrm{M})^{-1}\mathrm{A}^T b$$

yielding

$$\min_{\{u_k\}} \Delta J = -b^T\mathrm{A}(\mathrm{A}^T\mathrm{A}+M)^{-1}\mathrm{A}^T b$$

It follows that Eq. (4.17) holds if and only if b is not in the null space of A^T, i.e., if and only if

$$b^T\mathrm{A}\mathrm{A}^T b \neq 0$$

The regulability condition is then

$$\sum_{k=0}^{K-1}\left(\sum_{j=1}^{K-1}\mathrm{A}_{j,k}^T b_j\right)^T\left(\sum_{j=1}^{K-1}\mathrm{A}_{j,k}^T b_j\right) \neq 0$$

or, for any $x_0 \neq 0$

$$x_0^T\sum_{k=0}^{K-1}\left(\sum_{j=k}^{K-1}\Phi_{j+1,0}^T L_{j+1}\Phi_{j+1,k+1}\right)G_k G_k^T \cdot\left(\sum_{j=k}^{K-1}\Phi_{j+1,k+1}^T L_{j+1}\Phi_{j+1,0}\right)x_0 \neq 0$$

Noting that $\Phi_{j+1,0} = \Phi_{j+1,k+1}\Phi_{k+1,0}$, the condition becomes

$$x_0^T\sum_{k=0}^{K-1}\Phi_{k+1,0}^T\left(\sum_{j=k}^{K-1}\Phi_{j+1,k+1}^T L_{j+1}\Phi_{j+1,k+1}\right)G_k G_k^T \cdot\left(\sum_{j=k}^{K-1}\Phi_{j+1,k+1}^T L_{j+1}\Phi_{j+1,k+1}\right)\Phi_{k+1,0}x_0 \neq 0$$

for any $x_0 \neq 0$. The matrix

$$\psi_k = \sum_{j=k}^{K-1} \Phi_{j+1,k+1}^T L_{j+1} \Phi_{j+1,k+1}$$

can be seen to satisfy the (time reversed) Lyapunov equation

$$\psi_k = F_{k+1}^T \psi_{k+1} F_{k+1} + L_k, \quad \psi_k = L_k \tag{4.18}$$

The regulability condition can now be written as

$$x_0^T W_k x_0 \neq 0 \quad \text{for any } x_0 \neq 0$$

where W_k is the regulability Grammian

$$W_k = \sum_{k=0}^{K-1} \Phi_{k+1,0}^T S_k S_k^T \Phi_{k+1,0}$$

with

$$S_k = \psi_k G_k.$$

We have proved the following result:

Theorem 4.3. *The system Eq.* (4.1) *is regulable if and only if*

$$\text{rank } W_k = n \quad \text{for all } K \geq n \tag{4.19}$$

Δ

Suppose that the regularity condition is not satisfied. Then for some initial condition $x_0 \neq 0$, the cost will not be reduced by applying any input signal $\{u_k\}$. This initial condition can be found by solving the equation

$$\left(\sum_{k=0}^{K-1} \Phi_{k+1,0}^T S_k S_k^T \Phi_{k+1,0} \right) x_0 = 0$$

We note that the latter condition is analogous to the orthogonality condition

$$g^T E\{x_k y_j^T\} = 0, j \leq k$$

in the estimation problem.

4.3.1 *Time-invariant systems*

Now consider the time-invariant system

$$x_{k+1} = Fx_k + Gu_k \quad k \geq 0 \tag{4.20}$$

where F is assumed to have all its eigenvalues strictly inside the unit circle and the input $\{u_k\}$ is selected so as to minimize the cost function

$$J[x_0, \{u_k\}] = x_k^T L x_k + \sum_{k=0}^{K-1} x_k^T L x_k + u_k^T M u_k$$

For $K \to \infty$, the matrix ψ_k, defined by Eq. (4.18), has a steady-state value $\psi_k = \psi, k \geq 0$, which is the unique nonnegative solution of the equation

$$\psi = F^T \psi F + L \tag{4.21}$$

Taking $L = \psi$, we also have $\psi_k = \psi, k \geq 0$. This situation is the analog of stationarity of the stochastic system Eq. (4.1). Denoting $S = \psi G$, we obtain the following result as a direct consequence of Theorem 4.3.

Corollary 4.3. *The system Eq. (4.20) with $L = \Psi$ is regulable if and only if the pair (F^T, S) is reachable.*

Let f_i denote an eigenvector of F, corresponding to an eigenvalue λ_i. Multiplying Eq. (4.21) on the right by f_i, we obtain

$$\psi f_i = (I - \lambda F^T) L f_i$$

yielding

$$S^T f_i = G^T (\lambda_i^{-1} I - F^T) L f_i \tag{4.22}$$

By the PBH test, the pair (F^T, S) is reachable if and only if $S^T f_i \neq 0$ for any f_i corresponding to each λ_i. Noting that $\lambda_i \neq 0$, we obtain the following result from Corollary 4.1 and Eq. (4.22).

Corollary 4.4. *The system Eq. (4.20) with $L = \psi$ L is regulable if and only if for any eigenvector f_i corresponding to each eigenvalue λ_i of F,*

$$G^T (\lambda_i^{-1} I - F^T)^{-1} L f_i \neq 0 \tag{4.23}$$

It follows from Eq. (4.23) that a necessary condition for regulability is that $Lf_i \neq 0$ for any f_i corresponding to each λ_i, which is equivalent to reachability if the pair (F^T, L^T). Given that the latter condition is satisfied, a sufficient condition for regulability is that none of the zeros of the system (G, F^T, L) is a reciprocal of any of its poles.

The reachability of the pair (F^T, S), which, by Corollary 4.2, is the necessary and sufficient condition for regulability, was shown (Bernhard and Cohen, 1974) to be equivalent to the complete dependence of the regulation problem defined as observability of the pair (C, F), where C is the gain matrix in the optimal control law $u = -Cx$. This property was further shown (Bernhard and Cohen, 1974) to be equivalent to the nonreducibility of the Riccati equation associated with the optimal regulation problem. This relationship is analogous to the one between the estimability property and the reducibility of the Kalman filter, which is, in turn, equivalent to the reachability of (F, K), where K is the Kalman gain.

Finally, we note that, as in the estimation problem, our definition and result concerning regulability are readily extendable to the continuous-time case, where the regulability Grammian is defined in complete analogy to the discrete-time case and the term $(\lambda_i^{-1}I - F^T)$ is replaced by the term $(-\lambda_i I - F^T)$.

4.4 Duality

It can be seen that the estimability condition Eq. (4.1) with time-invariant parameters $\{F, G, H, Q, R\}$ and the time-invariant regulability condition Eq. (4.20), which involves the Lyapunov equations Eqs. (4.7) and (4.6), respectively, are similar, modulo time reversal. Such similarity, which has also been shown to exist between the Riccati equations associated with the estimation and the regulation problems, was called duality (Kalman, 1960). We now show that the duality between the estimability and the regulability properties is consistent with the one between the Lyapunov and the Riccati equations corresponding to the estimation and the regulation problems and, employing this duality, obtain the regulability condition from the estimability condition.

Estimability in the discrete-time case was defined in Section 4.2 by the condition

$$P_k < \Pi_k \tag{4.24}$$

where P_k is known to satisfy the Riccati equation

$$P_{k+1} = F_k P_k F_k^T + G_k Q_k G_k^T - F_k P_k H_k^T (H_k P_k H_k^T + R_k)^{-1} H_k P_k F_k^T \tag{4.25}$$

initialized at $P_0 = \Pi_0$, while Π_k satisfies the Lyapunov equation

$$\Pi_{k+1} = F_k \Pi_k F_k^T + G_k Q_k G_k^T \tag{4.26}$$

In the discrete-time regulation problem, the minimal cost is known to be (Kwakernaak and Sivan, 1972, p. 494).

$$\min_{\{u_k\}} J[x_0, \{u_k\}, K] = x_0^T \Gamma_0 x_0 \tag{4.27}$$

where Γ_0 satisfies the (time-reversed) Riccati equation

$$\begin{aligned} \Gamma_k &= F_{k+1}^T \Gamma_{k+1} F_{k+1} + L_{k+1} \\ &\quad - F_{k+1}^T \Gamma_{k+1} G_{k+1} \left[M_{k+1} + G_{k+1}^T \Gamma_{k+1} G_{k+1} \right]^{-1} G_{k+1}^T \Gamma_{k+1} F_{k+1} \end{aligned} \tag{4.28}$$

with the terminal condition $\Gamma_k = L_k$. The cost for zero input is obtained from Eqs. (4.1) and (4.2) as

$$J[x_0, \{u_k = 0\}, K] = x_0^T \psi_0 x_0 \tag{4.29}$$

where ψ_0 satisfies the (time-reversed) Lyapunov equation

$$\psi_k = F_{k+1}^T \psi_{k+1} F_{k+1} + L_{k+1} \tag{4.30}$$

with the terminal condition $\psi_k = L_k$. It follows from Eqs. (4.27) and (4.29) that the regulability condition Eq. (4.15) is equivalent to the condition

$$\Gamma_0 < \psi_0 \tag{4.31}$$

It can be seen that the regulability condition Eq. (4.31) and Eqs. (4.28) and (4.30) are dual, modulo time-reversal, to the estimability condition Eq. (4.24) and Eqs. (4.25) and (4.26). It follows that the regulability condition, dual to the estimability condition

$$\text{rank} \sum_{k=K-1}^{0} \Phi_{K,k}\Pi_k H_k H_k^T \Pi_k \Phi_{K,k}^T = n \quad \text{for all K} \geq n$$

is given by

$$\text{rank} \sum_{k=0}^{K-1} \theta_{0,k}\psi_k G_k G_k^T \Pi \psi_k \theta_{0,k} = n \quad \text{for all K} \geq n$$

where ψ_k is defined by Eq. (4.18) and $\theta_{0,k}$ is the transition matrix of the (adjoint) system $q_k = F_{k+1}^T q_{k+1}$. Since $\theta_{0,k} = F_1^T F_2^T \ldots F_k^T$, we have $\theta_{0,k} = \Phi_{k+1,0}^T$, yielding the regulability condition

$$\text{rank} \sum_{k=0}^{K-1} \Phi_{k+1,0}^T \Psi_k G_k G_k^T \Psi_k \Phi_{k+1,0} = n \quad \text{for all K} \geq n$$

which is the same as the condition Eq. (4.8) obtained for estimability.

4.5 Discussion

Estimability and regulability are properties of stochastic linear systems that determine the ability to reduce the mean-square error in state estimation and the quadratic cost in state feedback regulation strictly below their values when no observation or feedback is used. These properties constitute a mathematical foundation supporting the significance of feedback sensorimotor control addressed in later chapters of this book, where feedback control and sensory information are experimentally shown to improve natural and impaired movement well above the achievements of previously suggested open-loop approaches. Since regular walking is a sensorimotor sequence which can be mathematically characterized as a discrete-time linear control process, closed-loop feedback control, bringing the notions of estimability and regulability into play, is found to reduce the levels of the clinically measured biological sensory noise and control effort, substantially below their levels under open-loop control.

Neuron-based cortical models are essentially nonlinear. In contrast, the larger scales of sensorimotor functions have been represented by linear feedback control processes, e.g., (Nataraj and van den Bogert, 2017). In later chapters of this book, addressing more directly the biological and the medical domains, we note major advantages of closed-loop feedback control systems with respect to open-loop control systems. Such advantages are shown in the present and the next chapter, to be mathematically deduced from the notions of estimability and regulability. Specifically, these notions represent the abilities of closed-loop linear feedback control systems to improve the accuracies of state estimation and state regulation with respect to open-loop systems. It might be noted that the notions of estimability and regulability are independent of those of observability and controllability (Kalman, 1960; 1961) in the sense that one couple does not imply the other. In particular, observability and controllability, in contrast to the estimability and regulability, do not guarantee strict reduction in the posterior error covariance matrix or in the quadratic cost of optimal control.

Chapter 5

Minimal-Order Linear State Estimation Under Performance Constraints

5.1 Introduction

In certain applications, the implementation of linear state estimators for dynamical systems requires order reduction due to real-time computation and memory constraints. Under certain conditions on the measurement noise (Bryson and Johansen, 1965) and the internal system structure (Baram and Shaked, 1985), the order of the state estimator may be reduced without loss of accuracy with respect to the full-order Kalman filter. However, when these conditions do not hold, or when the required order reduction exceeds that allowed by the structure, performance degradation is to be expected. Clearly, the extent of degradation depends on the extent of order reduction. In actual application, the maximal allowable estimation inaccuracy must be imposed as a constraint on the order reduction procedure. Certain optimal mean-square solutions for time-varying systems, e.g., (Sims and Asher, 1978), have considered the case where only a part of the state vector is of interest. Others have allowed for reduced-order estimation of the full state vector, restricting themselves, however, to time-invariant systems at steady state (Bernstein and Hyland, 1987). An attempt to meet an accuracy specification using such techniques would require repeating the optimization procedure for different orders, which, for systems of high orders, may become a costly task. Lower and upper bounds on the minimal order needed for meeting the accuracy requirement can be useful in reducing the search set, thus reducing the implementation difficulties. Furthermore, such bounds may lead to simple minimal order estimators which satisfy the performance constraint, avoiding the higher computational effort required by direct parameter optimization.

In this chapter, we consider the reduced-order estimation of the full state vector of a time-varying linear system under a constraint on the maximal allowable mean-square error. We do not attempt to find a minimum mean-square error estimator by parameter optimization, as this approach is practically intractable for the problem at hand. Instead, we propose a bound-based approach (Baram and Kalit, 1987). We first derive a global lower bound on the mean-square error of any reduced-order estimator for the given system. The bound value at any time point and, in the case of a time-invariant system, at steady-state, is readily calculable, simultaneously for all reduced orders. A lower bound on the minimal estimator order, needed for meeting the accuracy constraint, readily follows. The attainability of the lower bound is then examined in two steps. First, assuming that the bound is attainable, the ideal reduced-order estimator, which satisfies the bound, is derived. Then, a necessary and sufficient condition for realizability of the ideal estimator is derived. When the ideal estimator is not realizable, its structure maintains, in some sense, a local optimality property with respect to the last observation and the last estimate. It defines a suboptimal reduced-order estimator, which we call the pseudo-ideal estimator. The mean-square error of the pseudo-ideal estimator constitutes an upper bound on the optimal error, which, in turn, defines an upper bound on the minimal estimator order, needed for meeting the accuracy requirement. The lower and the upper bounds on the order define the range of orders over which further parameter optimization may be performed. As illustrated by a practical numerical example, this reduced search set may become trivial and the ideal and the pseudo-ideal estimators may coincide in a specified numerical sense, eliminating the need for further parameter optimization.

5.2 A fundamental inequality

The following result is fundamental to the subsequent analysis. We use $E\{\cdot\}$ to denote the expectation operation and a.s. to denote "almost surely".

Theorem 5.1. *Let $x \in \mathrm{R}^n$ be a random vector on a probability space $(\bar{\Omega}, \bar{F}, p)$ and let $\hat{x} \in \mathrm{R}^n$ be a vector of linear functionals of a measurable field Y, such that $E\{(x - \hat{x}) f\} = 0$ for all linear functionals f of Y. Then, for any vector of linear functionals $\tilde{x} \in \mathrm{R}^n$ of Y, for which rank $E\{\tilde{x}\tilde{x}^T\} \leq r < n$, we have*

$$E\left\{(x - \tilde{x})^T (x - \tilde{x})\right\} \geq tr\ P + \sum_{i=r+1}^{n} \lambda_i \tag{5.1}$$

where

$$P = E\left\{(x - \hat{x})(x - \hat{x})^T\right\} \tag{5.2}$$

and λ_i denotes the i-th largest eigenvalue of

$$\Pi = E\left\{\hat{x}\hat{x}^T\right\} \tag{5.3}$$

with equality if and only if

$$\tilde{x} = T^r T^{rT} \hat{x} \ a.s. \tag{5.4}$$

where T^r is the matrix whose columns are the eigenvectors of Π corresponding to the r largest eigenvalues. Δ

5.3 Lower bounds on the mean-square error and on the minimal allowable order in reduced-order linear state estimation

Consider the system

$$x_{k+1} = A_k x_k + B_k w_k \tag{5.5a}$$

$$y_k = C_k x_k + v_k \tag{5.5b}$$

where $x_k \in \mathrm{R}^n$, $y_k \in \mathrm{R}^m$, $E\{x_0\} = E\{w_k\} = E\{v_k\} = 0$, $E\{x_0 x_0^T\} = P_0, E\{w_k w_i^T\} = Q\delta_{k,j}$, $E\{v_k v_i^T\} = R_k \delta_{k,i}$, where $\delta_{k,j}$ denotes the Kronecker delta, e.g., (Frankel, 2012), and $E\{w_k v_i^T\} = 0$, $E\{w_k v_i^T\} = 0$, $E\{w_k x_0^T\} = 0$, and $E\left\{v_k x_j^T\right\} = 0$. A state estimator of order r, where $r < n$, for the system Eq. (5.5) is a system of the form

$$\hat{x}^r_{k+1} = F_k \hat{x}^r_k + G_k y_k \quad \hat{x}^r_k \in R^r \tag{5.6}$$

from which a vector $\tilde{x}_k$ that approximates x_k in some sense, can be obtained by a linear transformation

$$\tilde{x}_k = H_k \hat{x}^r_k \tag{5.7}$$

where H_k is a matrix of dimension $n \times r$. Obviously, $\tilde{x}_k$ has a covariance of, at most, rank r. Hence, for any properly dimensioned H_k, we have

$$E\left\{(x_k - H_k \hat{x}^r_k)^T (x_k - H_k \hat{x}^r_k)\right\} \geq \operatorname{tr} P_k + \sum_{i=r+1}^{n} \lambda_i (\Pi_k) \equiv \varepsilon_{0,k}(r) \tag{5.8}$$

where the matrices

$$P_k = E\left\{(x_k - \hat{x}_k)(x_k - \hat{x}_k)^T\right\} \tag{5.9}$$

and

$$\Pi_k = E\left\{\hat{x}_k \hat{x}_k^T\right\} \tag{5.10}$$

are known to satisfy the equations (Anderson and Moore, 1979)

$$P_{k+1} = A_k P_k A_k^T + B_k Q_k B_k^T - A_k P_k C_k^T \left[C_k P_k C_k^T + R_k\right]^{-1} C_k P_k A_k^T \tag{5.11}$$

and

$$\Pi_{k+1} = A_k \Pi_k A_k^T + A_k P_k C_k^T \left[C_k P_k C_k^T + R_k\right]^{-1} C_k P_k A_k^T \tag{5.12}$$

with $\Pi_0 = P_0$ given. For the case where the system is time invariant, Π_k attains a steady-state value under a stability condition and P_k attains a steady-state value under stabilizability and detectability conditions (Anderson and Moore, 1979). Under these conditions, the lower bound on the reduced-order estimation error also attains, then, a steady-state value, which can be calculated from the steady-state versions of Eqs. (5.11) and (5.12).

It can be seen that the lower bound on the mean-square error is a monotone decreasing function of the order, which attains its minimal value of tr P_k at $r = n$. The error can be used, then, to determine a lower bound on the order, needed for maintaining the estimation error below a certain level. Specifically, if the maximal acceptable error is α_k, then a lower bound on the order is

$$r_0 = \min \left\{ r : \sum_{i=r+1}^{n} \lambda_{k,i} \leq \alpha_k - \text{tr } P_k \right\} \tag{5.13}$$

5.4 The ideal reduced-order estimator

It follows from Theorem 5.1 that any state estimate for the system Eq. (5.5) obtained from an estimator of reduced order r will have a mean-square error greater than that of

$$\tilde{x}_k = T_k^r x_k^r \tag{5.14}$$

where

$$x_k^r = T_k^{rT} \hat{x}_k \tag{5.15}$$

Since x_k^r is r-dimensional, any linear system which produces x_k^r from the observations is a minimum mean-square error state estimator of order r for the system Eq. (5.5). Since, as we shall see in the following section, such a system may not be realizable, we call it an ideal reduced-order estimator. We now derive an ideal reduced-order estimator under the assumption that it is realizable. It produces x_{k+1}^r from x_k^r and y_k by a linear transformation of the form

$$x_{k+1}^r = F_k x_k^r + G_k y_k \tag{5.16}$$

We shall find it convenient to replace the pair x_k^r, y_k by an equivalent uncorrelated pair. Let

$$C_k^r = E\left\{y_k x_k^{rT}\right\}\left[E\left\{x_k^r x_k^{rT}\right\}\right]^{-1} \tag{5.17}$$

then it can be readily verified that

$$v_k^r = y_k - C_k^r x_k^r \tag{5.18}$$

is orthogonal to x_k^r. Furthermore, since the pair x_k^r, y_k is related to the pair x_k^r, v_k^r by

$$\begin{bmatrix} x_k^r \\ y_k \end{bmatrix} = \begin{bmatrix} I & 0 \\ C_k^r & I \end{bmatrix} \begin{bmatrix} x_k^r \\ v_k^r \end{bmatrix} \tag{5.19}$$

they are equivalent for linear operations. An ideal estimator of order r for Eq. (5.5) then has the form

$$x_{k+1}^r = A_x^r x_k^r + K_k^r v_k^r \tag{5.20}$$

Multiplying Eq. (5.20) on the right by x_k^{rT} and taking expectation, we obtain

$$A_k^r = E\left\{x_{k+1}^r x_k^{rT}\right\} \left[E\left\{x_k^r x_k^{rT}\right\}\right]^{-1} \tag{5.21}$$

Multiplying Eq. (5.20) on the right by v_k^{rT} and taking expectation, we obtain

$$K_k^r = E\left\{x_{k+1}^r v_k^r\right\} \left[E\left\{v_k^r v_k^{rT}\right\}\right]^{-1} \tag{5.22}$$

We next derive the matrices A_k^r, C_k^r and K_k^r explicitly in terms of the given system Eq. (5.5). Let us denote by $\Lambda_k = \operatorname{diag}(\lambda_{k,1}, \dots, \lambda_{k,n})$ the diagonal matrix of the eigenvalues of Π_k arranged in descending order of magnitude and by T_k, the matrix whose columns are the corresponding eigenvectors of Π_k. Also let T_k^r denote the matrix composed of the first r columns of T_k and let $\Lambda_k^r = \operatorname{diag}(\lambda_{k,1}, \dots, \lambda_{k,r})$. Noting that

$$T_k^T \Pi_k T_k = \Lambda_k \tag{5.23}$$

which yields

$$\Pi_k T_k = T_k \Lambda_k \tag{5.24}$$

we obtain, after some manipulations

$$A_k^r = T_{k+1}^{rT} A_k T_k^r \tag{5.25}$$

$$C_k^r = C_k T_k^r \tag{5.26}$$

and

$$K_k^r = T_{k+1}^{tT} A_k P_k^r C_k^T \left[C_k^T P_k^r C_k + R_k \right]^{-1} \tag{5.27}$$

where

$$P_k^r = \Pi_k + P_k - T_k \tilde{\Lambda}_k^r T_k^r \tag{5.28}$$

with

$$\tilde{\Lambda}_k^r = \begin{bmatrix} \Lambda_k^r & 0 \\ 0 & 0 \end{bmatrix} \tag{5.29}$$

The state estimate for the system Eq. (5.5) is obtained from the ideal reduced-order estimator by Eq. (5.14). We note that the ideal reduced-order estimator derived above is unique in the sense that if $q_k \in R^r$ is the state of another reduced estimator and $\tilde{x}_k$ can be obtained from q_k by a linear transformation, then x_k^r and q_k are equivalent in the sense that they are related by linear transformations. Indeed if

$$\tilde{x}_k = H_k q_k$$

where H_k can be assumed, without loss of generality, to have a full rank, then

$$T_k^r x_k^r = H_k q_k$$

yielding

$$x_k^r = T_k^{rT} H_k q_k$$

and

$$q_k = \left(H_k^T H_k \right)^{-1} H_k^T T_k^r x_k^r$$

Combining Eqs. (5.14) and (5.15), we have

$$\tilde{x}_k = T_k^r T_k^{rT} \hat{x}_k$$

The estimation error covariance is

$$\begin{aligned}
E\left\{(x_k - \tilde{x}_k)(x_k - \tilde{x}_k)^T\right\} &= P_k + E\left\{(\hat{x}_k - \tilde{x}_k)(\hat{x}_k - \tilde{x}_k)^T\right\} \\
&= P_k + \left(I - T_k^r T_k^{rT}\right)\Pi_k\left(I - T_k^r T_k^{rT}\right) \\
&= P_k + \Pi_k - \Pi_k T_k^r T_k^{rT} - T_k^r T_k^{rT}\Pi_k \\
&\quad + T_k^r T_k^{rT}\Pi_k T_k^r T_k^{rT} \\
&= P_k + \Pi_k - T_k \tilde{\Lambda}_k^r T_k^T \\
&= P_k + \Pi_k - N_k \\
&= P_k^r
\end{aligned} \tag{5.30}$$

It can be seen that for $r = n$, we have $P_k^r = P_k$, as should be expected. For $r = 0$, i.e., no measurements incorporated in the estimation, we have $P_k^r = P_k + \Pi_k = \Psi_k$, again, as should be expected. For $0 < r < n$, the penalty for order reduction is the increase in the error covariance

$$\Pi_k - N_k = T_k \Lambda_k T_k^T - T_k \tilde{\Lambda}_k^r T_k^T = T_k \bar{\Lambda}_k^r T_k^T \tag{5.31}$$

where

$$\bar{\Lambda}_k^r = \begin{bmatrix} 0 & & & & & \\ & \ddots & & 0 & & \\ & & 0 & & & \\ & & & \lambda_{k,r+1} & & \\ & & & & \ddots & \\ & 0 & & & & \lambda_{k,n} \end{bmatrix}$$

Finally, it should be noted that the gain of the ideal estimator of reduced order r has a form similar to that of the familiar Kalman gain, multiplied by T_{k+1}^{rT}, with the optimal estimation error covariance P_k replaced by the ideal reduced-order estimation error covariance P_k^r.

lity of the ideal reduced-order state

r

In the prev... section, we derived the ideal reduced-order estimator of order r, i.e., the one which satisfies the lower bound derived in Section 5.3, assuming that such an estimator exists. We now examine the realizability of the ideal reduced-order estimator, i.e., the attainability of the lower bound. The ideal reduced-order state estimator for the system Eq. (5.5) is said to be realizable if the minimum mean-square error estimate of dimension r, x_k can be produced by a linear system. We have seen that if this is the case, the ideal reduced-order estimator is specified by Eq. (5.20). We have the following result.

Theorem 5.2. *The ideal reduced-order estimator of order r for the system Eq.* (5.20) *is realizable if and only if the following conditions hold simultaneously for each k:*

$$\text{i) } K_k^r = T_{k+1}^{rT} K_k \tag{5.32}$$

$$\text{ii) } \Theta_k \Lambda_k^{-r} = 0 \tag{5.33}$$

where

$$K_k = \Lambda_k P_k C_k^T \left(C_k P_k C_k^T + R_k \right)^{-1} \tag{5.34}$$

$$\Lambda_k^{-r} = \begin{bmatrix} \lambda_{k,r+1} & & 0 \\ & \ddots & \\ 0 & & \lambda_{k,n} \end{bmatrix} \tag{5.35}$$

and Θ_k is the $r \times (n-r)$ dimensional upper right portion of the matrix $T_{k+1}^T \Gamma_k T_k$, where

$$\Gamma_k = A_k - K_k C_k \tag{5.36}$$

Δ

As a direct consequence of Theorem 5.2, we have the following result.

Corollary 5.1. *Let*

$$l = \max(0,\ rank\,\Pi_k - r)$$

then if

$$l = 0$$

the ideal estimator of order r for Eq. (5.5) *is realizable if and only if*

$$\theta_k^i = 0 \quad i = 1, \ldots, l \tag{5.37}$$

where θ_k^i *is the i-th column of* Θ_k. △

It is not difficult to see that the ideal estimator of order r is equivalent to the Kalman filter if and only if

$$\text{rank}\ \Pi_k \leq r \tag{5.38}$$

This follows from the fact that while by definition

$$x_k^r = T_k^{rT} \hat{x}_k$$

we also have in this case the inverse transformation

$$\hat{x}_k = T_k^r$$

Consequently, when rank $\Pi_k > r$, the ideal estimator of order r is not equivalent to the Kalman filter. This does not necessarily imply, however, that the ideal reduced-order estimator is not realizable, as can be seen from the following example.

Example 5.1. Let P_0 be diagonal, $B_k = C_k = I$ and A_k, Q_k and R_k be diagonal for all k. Then, clearly, Π_k, P_k and K_k are diagonal, $T_k = I$ and $T_k^{rT} = [I_r\ 0]$. In this case, we have

$$P_k^r = P_k + \bar{\Lambda}_k^r$$

hence, since $C_k = I$,

$$C_k P_k^r C_k^T + R_k = C_k P_k C_k^T + R_k + \bar{\Lambda}_k^r$$

It follows that $\left(C_k P_k^r C_k^T + R_k\right)^{-1}$ and $\left(C_k P_k C_k^T + R_k\right)^{-1}$ differ only in their $(n-r) \times (n-r)$ dimensional lower right portions, and, consequently, so do $\Lambda_k P_k^r C_k^T \left(C_k P_k^r C_k^T + R\right)^{-1}$ and $A_k P_k C_k^T \left(C_k P_k^r C_k^T + R_k\right)^{-1}$. But since $T_{k+1}^{rT} = [I_r\ 0]$, it follows that

$$K_k^r = T_{k+1}^{rT} K_k$$

Condition (i) of Theorem 5.2 is thus satisfied. Since for the case under consideration, Γ_k is diagonal, it follows that condition (ii) of Theorem 5.2 is also satisfied. It follows then from Theorem 5.2 that the ideal estimator of order r is realizable, independently of rank Π_k. (It is not difficult to see that this estimator is the upper r portion of the Kalman filter.)

We have seen that the ideal estimator of order r exists under certain structural conditions. There is no reason, however, to expect the estimator to exist in general. In fact, an examination of the conditions of Theorem 5.2 indicates that the class of systems for which the ideal estimator of order r exists is highly restrictive. Yet, from a practical viewpoint, the realizability question may be examined in terms of approximation. To this end, we next define a suboptimal, realizable version of the ideal estimator.

5.6 The pseudo-ideal reduced-order estimator: Upper bounds on the optimal error and the minimal order

When the conditions of Theorem 5.1 do not hold, the lower bound presented in Section 5.3 (Eq. (5.13)) is not attainable by a linear estimator. In other words, the ideal reduced-order estimator is not realizable. However, as shown below, the structure of the ideal estimator, which is, in a certain sense, locally optimal with respect to the last observation and the last estimate, constitutes a reduced-order estimator, which we call pseudo-ideal.

The ideal estimator of order r produces x_{k+1}^r from x_k^r and y_k by a linear transformation. When such a transformation does not exist, the ideal estimator represents $\hat{x}_{k+1}^r$, the best linear projection in the minimum mean-square error sense, of x^r_{k+1} on x^r_k and y_k, i.e.,

$$\hat{x}^r_{k+1} = \left(A^r_k - K^r_k C^r_k\right) x^r_k + K^r_k y_k \tag{5.39}$$

where A^r_k, C^r_k and K^r_k are defined by Eqs. (5.25), (5.26) and (5.27), respectively.

Unfortunately, x^r_k is not available. Instead, we substitute its estimate $\hat{x}^r_k$. This results in a reduced-order estimator

$$\hat{x}^r_{k+1} = \left(A^r_k - K^r_k C^r_k\right) \hat{x}^r_k + K^r_k y_k \tag{5.40}$$

having the same parameters as the ideal reduced-order estimator. This is the pseudo-ideal reduced-order estimator. The state estimate for Eq. (5.6) is obtained from the pseudo-ideal estimator as

$$\tilde{x}_k = T^r_k \hat{x}^r_k$$

The associated mean-square error

$$\varepsilon^0_k(r) = E\left\{\left(x_k - \tilde{x}_k\right)^T \left(x_k - \tilde{x}_k\right)\right\} \tag{5.41}$$

can be obtained as

$$\varepsilon^0_k(r) = \operatorname{tr} \mathrm{M}_k \tag{5.42}$$

where

$$M_k = E\left\{\left(x_k - \tilde{x}_k\right)\left(x_k - \tilde{x}_k\right)^T\right\}$$

is calculated as

$$M_k = \left[I \quad -T^r_k\right] \psi_k \begin{bmatrix} I \\ -T^{rT}_k \end{bmatrix} \tag{5.43}$$

where

$$\psi_k = E\left\{\begin{bmatrix} x_k \\ \hat{x}^r_k \end{bmatrix} \begin{bmatrix} x^T_k & \hat{x}^{rT}_k \end{bmatrix}\right\} \tag{5.44}$$

satisfies the equation

$$\psi_{k+1} = A_k \psi_k A^T_k + Q_k \tag{5.45}$$

where

$$\bar{A}_k = \begin{bmatrix} A_k & 0 \\ K_k^r C_k & A_k^r C_k^r \end{bmatrix}$$

$$Q_k = \begin{bmatrix} B_k Q_k B_k^T & B_k S_k K_k^{sT} \\ K_k^r S_k^T B_k^T & K_k^r R_k K_k^{sT} \end{bmatrix}$$

and

$$\psi_0 = \begin{bmatrix} P_0 & P_0 T_0^{rT} \\ T_0^r P_0 & T_0^r P_0 T_0^{rT} \end{bmatrix}$$

The mean-square error $\varepsilon_k^0(r)$ of the pseudo-ideal estimator constitutes an upper bound on the optimal mean-square error. An upper bound on the order of the optimal reduced-order estimator, needed to satisfy a specified constraint on the mean-square error, α_k, is defined by

$$r^0 = \min\left\{r : \varepsilon_k^0(r) \leq \alpha_k\right\} \tag{5.46}$$

It should be noted that similar upper bounds can be obtained using any other suboptimal reduced-order estimator. The advantage in using the pseudo-ideal estimator is that it becomes the ideal, optimal estimator when the conditions of Theorem 5.2 are satisfied. As explained and illustrated in the following section, while these conditions may not hold precisely, the difference between the ideal and the pseudo-ideal estimator should be examined on a numerical basis, and, in application, the two estimators may be sufficiently close so as to be considered practically the same.

5.7 Search set reduction in minimal-order estimator design under performance constraints

A state estimator of minimal order which satisfies a constraint on the mean-square error may be found, in principle, by repeated parameter optimization for different orders. As noted before, parameter optimization may not be tractable in nonstationary cases, and suboptimal alternatives must be considered. However, even when parameter optimization is practically realizable, as in the stationary case (Bernstein and Hyland, 1985), repeated optimization for various orders may be costly, and reduction of the search set may be desirable. Such reduction is offered by the lower bound represented by Eq. (5.13), which is easy to calculate for both stationary and nonstationary systems and reduces the search to the set $R_1 = \{r \geq r_0\}$. A further reduction of the search set is offered by the upper bound in Eq. (5.46), which, together with the lower bound, defines the set

$$R = \left\{r_0 \leq r \leq r^0\right\}.$$

Clearly, the pseudo-ideal estimator of order r^0 satisfies the constraint. However, order reduction below r^0 may be attempted by further search within R. When the ideal and the pseudo-ideal estimators are sufficiently close, so that the set R consists of a single r value, the design problem is solved without the need for further search. Even if this is not the case, the set R can be expected to be considerably smaller than the original search set, leading to a considerable reduction in the computational requirement of the design procedure.

The size of the set R depends on the required performance level and on the difference between the ideal and the pseudo-ideal estimators. This difference represents the extent to which the ideal estimator is not realizable. It can be measured by

$$d(r,k) = \frac{\varepsilon_k^0(r) - \varepsilon_{0,k}(r)}{\varepsilon_k^0(r)}$$

where the error bounds $\varepsilon_{0,k}(r)$ and $\varepsilon_k^0(r)$ are defined by Eq. (5.8) and Eqs. (5.41)–(5.45), respectively. The ideal estimator may be considered realizable, i.e., identical to the pseudo-ideal estimator if $d(r,k) \leq 1$. When R consists of a single r value, i.e., when $r_0 = r^0 \equiv r^*$, the stated design problem is solved, as the pseudo-ideal estimator of order r^* is an estimator of minimal order which satisfies the constraint. Yet, further reduction of the mean-square error by parameter optimization may be attempted. This, however, is not necessary if $d(r^*,k) \ll 1$, as then, for all practical purposes, the ideal, the pseudo-ideal, and the parameter-optimal estimators coincide. These considerations are well illustrated by the following example.

Example: A seventh-order linear model for a power system (Mahalanabis and Ray, 1986) has been discretized at sampling intervals of 0.05 s, and the steady-state values of the matrices $P, \Pi,$ and T have been calculated. The eigenvalues of Π were found to be

$$\lambda_1 = 0.74886,\ \lambda_2 = 0.16036,\ \lambda_3 = 0.083515,\ \lambda_4 = 0.0092170,$$
$$\lambda_5 = 0.0016823, \lambda_6 = 0.000056759,\ \lambda_7 = 0.000054829$$

while

$$\operatorname{tr} P = 0.0632$$

The lower bound on the mean-square error of any estimator of order r was obtained from

$$\varepsilon^2 = \operatorname{tr} P + \sum_{i=r+1}^{7} \lambda_i$$

for $r = 1, 2, \ldots, 7$. The results are shown in Fig. 5.1.

The mean-square error of the pseudo-ideal reduced-order estimator of Eq. (5.40) in steady state has been calculated for $r = 1, \ldots, 7$, and its values are shown in Fig. 5.1. It can be seen that, for $r \geq 3$, the mean-square error of the pseudo-ideal estimator is close to that of the ideal, i.e., to the lower bound, which implies that the pseudo-ideal estimator, which is easy to derive, can be used instead of the optimal reduced-order estimator with practically the same accuracy. In fact, for $r \geq 4$, the reduced-order estimator can be used instead of the full-order Kalman filter, with practically no loss of accuracy.

It can be seen from Fig. 5.1 that, for a maximal allowable mean-square error of 0.15, any estimator of order 2 or less will violate the lower bound. The pseudo-ideal estimator of order 3 satisfies the constraint and is, therefore, an acceptable solution to the problem (it is an estimator of minimal order, which satisfies the constraint). If the maximal allowable order is increased to, say, 0.3, one may attempt to reduce the estimator order to 2 by optimizing the parameters of an estimator of order 2, using an earlier technique (e.g., Bernstein and

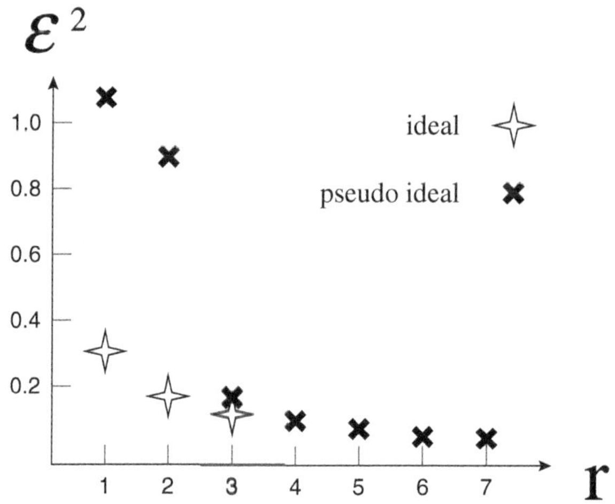

Figure 5.1. The mean-square errors of the ideal and the pseudo-ideal reduced-order estimators.

Hyland, 1985) and checking whether the resulting estimator satisfies the constraint.

We emphasize in closing that the above considerations apply, with similar simplicity, to time-varying systems.

5.8 Conclusion

A bound-based approach to the design of minimal-order linear state estimators under performance constraints has been presented. Lower bounds on the optimal mean-square error and on the minimal estimator order were derived, along with an ideal reduced-order estimator, which satisfies the bounds subject to realizability conditions. It has been shown that when the ideal estimator is not realizable, its structure forms a practical reduced-order estimator, namely, the pseudo-ideal estimator, which maintains, in some sense, a local optimality criterion. The mean-square error of the pseudo-ideal estimator can be used to upper-bound the optimal error and the optimal estimator order. The lower and the upper bounds on the order reduce the search set for the optimal, minimal order estimator and, sometimes, eliminate the parameter optimization problem altogether, as illustrated by a numerical example.

Chapter 6

Minimal-Order Model Identification

6.1 Introduction

The minimal order representation of the relationship between the input and the output of a finite dimensional linear system has long been of interest (Kalman, 1963; Gilbert, 1963). Construction of such representations from a finite dimensional Hankel matrix (a square matrix in which each ascending skew-diagonal from left to right is constant) of impulse responses has been proposed (Ho and Kalman, 1966; Rissanen, 1971; Silverman, 1971, among others). In the case where the input is a stochastic process, the impulse response function is naturally replaced by the cross-correlation function between the input and the output. Probabilistic and geometric interpretations have been given to the role played by the Hankel correlation matrix when the input is a white noise process (Akaike, 1974a; 1974b).

When the cross-correlation function between the input and the output of the system under consideration is not given, one is faced with an identification problem. When the exact, unknown correlations are replaced by their estimates from measurement data, the finite structure of the Hankel correlation matrix is normally lost due to the inaccuracies associated with finite samples. A major consequence is that the minimal representation order cannot be determined exactly from any finite dimensional Hankel matrix of sample correlations. In the past, direct identification of the order from a Hankel matrix of sample correlations has been essentially given up due to the complexity of the probability distributions involved and other, indirect methods have been proposed by (Woodside, 1970; Chow, 1972a; 1972b). Order estimation criteria have been suggested and shown to produce consistent estimates (Akaike, 1974a; 1974b; Fine and Hwang, 1979). An approximation technique based on canonical variates analysis of truncated data records has also been suggested (Larimore, 1983).

Following our previously published work (Baram and Porat, 1988), this chapter discusses the problem of identifying a minimal order state space representation of a multivariable linear time-invariant system from Gaussian stationary input-output measurements. A procedure for identifying the system's order is proposed, based on an approximate probability distribution of the squared singular values of the Hankel matrix built from the sample cross-covariances. The approximate distribution converges to the true one as the number of measurements becomes large. The order determination procedure also identifies sets of linearly independent rows and linearly independent columns of the Hankel correlation matrix, which form a basis for a minimal order representation of the system.

6.2 Minimal-order system representation

Let $u_n \in \mathrm{R}^n$ be an uncorrelated Gaussian input of a system and let $y_n \in \mathrm{R}^m$ be the system's output. Suppose that u_n and y_n are jointly Gaussian and stationary with a cross-correlation function given by

$$R_k = E\left\{y_n u_{n-k}^T\right\} \tag{6.1}$$

where E denotes the expectation operation. We assume that there exist a positive integer P and a set of positive scalars $a_1, \ldots, a_{p-1}, a_0 = 1$, such that, for any $k \geq p$

$$\sum_{i=0}^{P-1} a_i R_{k-i} = 0 \tag{6.2}$$

It is well known that when u_n is the input and y_n is the output of a finite dimensional linear system, they satisfy a relationship of the type Eq. (6.2), where the a_i's are the system's characteristic polynomial coefficients. In the context of system identification, u_n may represent a known, synthetically generated "pseudo-random" sequence.

It is desired to find a minimal order representation for the input-output relationship between u_n and y_n. Let us denote, by $U_n^- = \left(u_{n-1}^T, u_{n-2}^T, \ldots, u_0^T\right)^T$, the vector of past inputs and, by $Y_n^+ = \left(y_n^T, y_{n+1}^T, \ldots, y_{2n-1}^T\right)^T$, the vector of n-step future outputs. The space $Y_n^+ \left| U_n^-\right.$ of mean-square projection of Y_n^+ on U_n^- is spanned by $R(n)\left[E\left\{U_n^- U_n^{-T}\right\}\right]^{-1} U_n^-$, where

$$R(n) = \begin{bmatrix} R_1 & R_2 & \cdots & R_n \\ R_2 & R_3 & \cdots & R_{n+1} \\ \vdots & & & \\ R_n & R_{n+1} & \cdots & R_{2n-1} \end{bmatrix} \tag{6.3}$$

It follows from Eq. (6.2) that the rows beyond the P-th block row of $R(n)$ are linearly dependent on the previous ones. This implies that $Y_n^+ \left| U_n^-\right.$ is spanned by $R\left[E\left\{U_n^- U_n^{-T}\right\}\right]^{-1} U_n^-$, where

$$R(P,n) = \begin{bmatrix} R_1 & R_2 & \cdots & R_n \\ R_2 & R_3 & \cdots & R_{n+1} \\ \vdots & & & \\ R_P & R_{P+1} & \cdots & R_{P+n-1} \end{bmatrix} \tag{6.4}$$

Suppose that $R(P,n)$ has p linearly independent rows and let x_n denote the vector whose elements are the Euclidean inner products between some maximal set of linearly independent rows of $R(P,n)$ and $\left[E\left\{U_n^- U_n^{-T}\right\}\right]^{-1} U_n^-$. Then x_n is a vector of minimal dimension, which spans $Y_n^+ \left| U_n^-\right.$. It follows that there exist matrices A_n, B_n, C_n and D_n such that

$$x_{n+1} = A_n x_n + B_n u_n, \quad y_n = C_n x_n + D_n u_n \tag{6.5}$$

Let $U_n^-(P) = \left(u_{n-1}^T, u_{n-2}^T, \ldots, u_{n-p}^T\right) T$ denote the P-step past of u_n and let $Y_n^+(P) = \left(y_n^T, y_{n+1}^T, \ldots, y_{n+P-1}^T\right)^T$ denote the P-step future of y_n. Let us denote

$$R = R(P) = E\left\{Y_n^+(P)U_n^-(P)^T\right\} = \begin{bmatrix} R_1 & R_2 & \cdots & R_P \\ R_2 & R_3 & \cdots & R_{P+1} \\ \vdots & & & \\ R_P & R_{P+1} & \cdots & R_{2P-1} \end{bmatrix}$$

Since the columns beyond the P-th block column of $R(P,n)$ are linearly dependent on the previous ones, R is of rank p. Since we have

$$E\left\{E\left\{Y_n^+(P)\left|U_n^-\right.\right\}U_n^-(P)^T\right\} = E\left\{Y_n^+(P)U_n^-(P)^T\right\}$$

and since x_n consists of elements of $E\left\{Y_n^+(P)\left|U_n^-\right.\right\}$, there exists a p-dimensional vector ω_n of elements of $U_n^-(P)$, such that the matrix $E\left\{x_n\omega_n^T\right\}$ is of full rank. By simple operations on Eq. (6.5), one obtains

$$A_n = E\left\{x_{n+1}\omega_n^T\right\}\left[E\left\{x_n\omega_n^T\right\}\right]^{-1} \tag{6.6a}$$

$$B_n = E\left\{x_{n+1}u_n^T\right\}\left[E\left\{u_n u_n^T\right\}\right]^{-1} \tag{6.6b}$$

$$C_n = E\left\{y_n\omega_n^T\right\}\left[E\left\{x_n\omega_n^T\right\}\right]^{-1} \tag{6.6c}$$

$$D_n = E\left\{y_n u_n^T\right\}\left[E\left\{u_n u_n^T\right\}\right]^{-1} \tag{6.6d}$$

This is, essentially, Akaike's representation (Akaike, 1974a; 1974b). We note that the term $D_n u_n$ is missing in Akaike's representation due to inclusion of u_n in U_n^-. Here, we have chosen to obtain the form of Eq. (6.5), which is often encountered in stochastic system estimation and control.

We proceed to obtain explicit expressions for the matrices A_n, B_n, C_n and D_n. Let W denote the submatrix of R, which consists of the intersection between the first maximal set of linearly independent rows and an equal number of the first linearly independent columns of R, i.e., the elements of R, which belong to both sets. Let

$$S == \begin{bmatrix} R_2 & R_3 & \cdots & R_{P+1} \\ R_3 & R_4 & \cdots & R_{P+2} \\ \vdots & & & \\ R_{P+1} & R_{P+2} & \cdots & R_{2P} \end{bmatrix}$$

and denote by U the matrix obtained by intersecting the rows and columns of S having the same indices as the rows and the columns of W in R. Then, by the arguments leading to Eq. (6.6), we obtain

$$A_n = A = UW^{-1} \tag{6.7a}$$

Denote by V the matrix, which consists of the intersection between the first q columns of R with its first p linearly independent rows; then

$$B_n = B = V\left[\text{cov}\left(u_n\right)\right]^{-1} \tag{6.7b}$$

Let Z denote the matrix, which consists of the intersection of the first block row of R with the columns of W. Then

$$C_n = C = ZW^{-1} \tag{6.7c}$$

Finally,

$$D_n = D = R_0\left[\text{cov}\left(u_n\right)\right]^{-1} \tag{6.7d}$$

A minimal order representation for the input-output relationship between u_n and y_n is then given by

$$x_{n+1} = Ax_n + Bu_n, \quad y_n = Cx_n + Du_n$$

where A, B, C and D are given by Eqs. (6.7a)–(6.7d). We note that due to the finite initial input time, $n = 0$, imposed in the definition of U_n^-, the state x_n, as defined above, is nonstationary. Passing to initialization in the remote past, x, becomes stationary, with $\Pi = E\left\{x_n x_n^T\right\}$ satisfying

$$\Pi = A\Pi A^T + B\text{cov}\left(u_n\right)B^T$$

We have seen that the key to the construction of a minimal order representation for the input-output relationship between the processes u_n and y_n from their cross-correlation function is the construction of the matrix W, i.e., the selection of maximal sets of linearly independent rows and linearly independent columns of the matrix R. The following procedure, based on a technique for partial realization from impulse responses (Kung, 1977), selects a maximal set of linearly independent columns of R. Modifying the technique so as to select linearly independent rows of R instead of linearly independent columns is straightforward. Define

$$\tilde{R} = R(2P) = \begin{bmatrix} R_1 & R_2 & \cdots & R_{2P} \\ R_2 & R_3 & \cdots & R_{2P+1} \\ \vdots & & & \\ R_{2P} & R_{2P+1} & \cdots & R_{4P-1} \end{bmatrix}$$

Set $j = 0$ and $i = 1$. Check the rank of the matrix $H_{j,i}$, composed of the upper left $(2P - j) \times (jm + i)$ submatrix of $\tilde{R}$. If $H_{j,i}$ is rank deficient, delete the $(km + i)$-th columns, $k = j, j + 1, \ldots, P$ from $\tilde{R}$ and repeat the rank check for the new $H_{j,i}$. If $H_{j,i}$ has a full rank and $i < m$, set $i = i + 1$ and repeat the rank check. If $i = m$, set $j = j + 1$ and $i = 1$. This procedure is continued until rank $H_{j,i} = P$ or until $j = Pm$. The rank p of R is the rank of the final $H_{j,i}$. A maximal set of the first linearly independent columns of R is obtained by taking the first p elements of the columns of $H_{j,i}$.

It can be seen that the above procedure guarantees that each tested matrix has, at most, rank deficiency of order one, i.e.,

$$\operatorname{rank} H_{i,j} \geq (jm + i - 1) \tag{6.8}$$

(clearly, rank $H_{j,i} \leq jm + i$). This property is consistent with the statistical inference method discussed in the following section.

6.3 Statistical inference

We now turn to the problem of finding a minimal order representation for a vector-valued process when the correlations R_i are not given. Instead, we assume that estimates $\hat{R}_i$ are available. Specifically, we take $\hat{R}_i$ to be the sample covariances. A natural approach to this problem is substituting the values of $\hat{R}_i$ instead of R_i in the representation derived in Section 6.2. The problem is, however, that the minimal order is unknown and $\hat{R}$ is likely to be of full rank for any P due to the stochastic nature of the problem. The decision on the order and on the linear independence of rows and of columns of R, which form the basis for the minimal representation, must be treated, then, by statistical means. Following the procedures suggested in the preceding section, the order and a maximal set of linearly independent rows and linearly independent columns will be selected simultaneously by performing sequential tests on the ranks of the matrices H_i. In the sequel, we derive an approximate distribution for such tests.

Let H represent the matrix H_i, which is to be checked for having a full rank in the procedure described in the previous section. Let us denote by $\lambda_i = \lambda_i(H)$, $i = 1, 2, \ldots$, the eigenvalues of $H^T H$ (or the squared singular values of H) in descending order. Suppose that the smaller dimension of H is k (i.e., that H has k singular values). Clearly, H has a full rank if and only if λ_k is nonzero. A test on H having a full rank can then be formulated as a test on λ_k having a nonzero value. Since our derivation of the test distribution will require the assumption that λ_{k-1} is nonzero when testing λ_k, and since this assumption is consistent with our rank checking procedure, we shall make it implicitly when necessary.

Let r denote a vector of the distinct scalar elements of $R(P)$ arranged in some order. Let $\hat{r}$ denote an estimate of r and denote by $\hat{\lambda}_i = \lambda_i(\hat{H})$, $i = 1, \ldots, k$ the eigenvalues of $\hat{H}^T \hat{H}$, where the latter is obtained from $H^T H$ by substituting in it the values $\hat{R}_j$ instead of R_j. The second-order Taylor series expansion of $\hat{\lambda}_k$ about r is

$$\hat{\lambda}_k = \lambda_k + f^T \tilde{r} + \frac{1}{2} \tilde{r}^T F \tilde{r} \tag{6.9}$$

where

$$\tilde{r} = \hat{r} - r, \quad f = \frac{\partial \lambda_k}{\partial r}$$

and

$$F_{i,j} = \frac{\partial^2 \lambda_k}{\partial r_i \partial r_j}$$

where r_i is the i-th element of r. As is well known, e.g., (Kung, 1977, p. 228, Thm. 14), the sample covariance $\hat{R}_i$ is asymptotically normally distributed with mean R_i and covariance inversely proportional to the number of measurements T. Therefore, the components of $\tilde{r}$ are asymptotically of the order of $1/\sqrt{T}$. This justifies the second-order Taylor series approximation given above for large enough T. We note that under the tested hypothesis we have $\lambda_k = 0$. Since $H^T H \geq 0, \lambda_k = 0$ is a global minimum. Hence, we have $f = 0$, so Eq. (6.9) assumes the form

$$\hat{\lambda}_k = \frac{1}{2}\tilde{r}^T F \tilde{r} \tag{6.10}$$

The vector $\tilde{r}$ is asymptotically distributed as a normal random vector. Let us denote by Σ the covariance of $\tilde{r}$ and by Θ the matrix of eigenvectors of $\Sigma^{1/2}$, then

$$\Theta^T \Sigma^{1/2} F \Sigma^{1/2} \Theta = L$$

where L is a diagonal matrix whose elements are denoted by α_i, $i = 1, \ldots, n$, where n is the dimension of r. Defining

$$z = \Theta \Sigma^{-1/2} \tilde{r}$$

we can see that z is asymptotically distributed as a normal vector with mean zero and variance I_n.

The quadratic form Eq. (6.10) can now be written as

$$\hat{\lambda}_k = \sum_{i=1}^{n} a_i z_i^2$$

where $z_i^2, i = 1, \ldots, n$ are independent chi-squared random variables with one degree of freedom. The characteristic function of $\hat{\lambda}_k$ is obtained as, e.g., (Hannan, 1970)

$$\Phi(\omega) = \prod_{j=1}^{n} \left(1 - 2i\alpha_j \omega^{-1/2}\right)$$

The distribution of $\hat{\lambda}_k$ can be found by integrating $\Phi(\omega)$. Several integration and approximation techniques have been suggested, e.g., (Imhoff, 1961; Hannan, 1970; Johnson and Kotz, 1970). Given a confidence level (i.e., an acceptable probability of making the right decision), the resulting distribution defines a statistical test on the rank of the matrix H. A sequence of such tests, performed in the order of the procedure described in Section 6.2, will provide estimates of the minimal order and maximal sets of linearly independent rows and linearly independent columns of R.

It follows from the above derivation that statistical inference on the order and the linearly independent rows of R requires knowledge of F and E. These are derived in the following section and in this chapter's Appendix as functions of the theoretical correlations R_i. In practice, however, the values of R_i are unknown and will be replaced by estimates $\hat{R}_i$. Once maximal sets of linearly independent rows and linearly independent columns of R_i have been estimated, an estimate of the values presented in Section 6.2 may be obtained by substituting the estimated correlations in the corresponding matrices of Eqs. (6.7a)–(6.7d).

6.4 Deriving the rank-test statistics

We now seek the second derivative of the eigenvalues of the matrix $H^T H$ with respect to the vector r, as defined in the previous section. Let

$$H^T H = Q\Lambda Q^T \tag{6.11}$$

be the eigenvalue/eigenvector decomposition of the symmetric matrix $H^T H$. The diagonal elements of Λ will be denoted $\{\lambda_1, \lambda_2, \ldots, \lambda_n\}$, while the columns of Q will be denoted $\{q_1, q_2, \ldots, q_n\}$. All the eigenvalues are assumed to be distinct. Let r_i, r_j denote any two components of the vector r (which may be identical or distinct). We wish to compute

$$\left\{\frac{\partial \lambda_l}{\partial r_i}; \frac{\partial^2 \lambda_l}{\partial r_i \partial r_j}; 1 \leq l \leq k\right\} \tag{6.12}$$

In particular, we are interested in these derivatives for $l = k$, in the special case where $\lambda_k = 0$ and $\{\lambda_1, \lambda_2, \ldots, \lambda_{k-1}\}$ are nonzero. As we will see, the computation of the second derivatives requires the computation of the derivatives

$$\left\{\frac{\partial q_l}{\partial r_i}; 1 \leq l \leq k\right\}$$

6.4.1 *The first derivative of the eigenvalues of $H^T H$*

Let us denote for convenience

$$M(\lambda) = \lambda I - H^T H = Q\left[\lambda I - \Lambda\right] Q^T$$

By the definition of the eigenvalues, we have

$$\det\left(M(\lambda)\right)|_{\lambda=\lambda_i} = 0$$

Differentiation yields

$$\begin{aligned}\frac{\partial}{\partial r_i}\det\left(M\left(\lambda_i\right)\right) &= \operatorname{tr}\left\{\operatorname{adj}\left(M\left(\lambda_i\right)\right)\right\}\frac{\partial M\left(\lambda_i\right)}{\partial r_i}\\ &= \operatorname{tr}\left\{\operatorname{adj}\left(M\left(\lambda_i\right)\right)\left(\frac{\partial \lambda_i}{\partial r_i}I - \frac{\partial}{\partial r_i}H^T H\right)\right\} = 0\end{aligned}$$

where $\operatorname{tr}\{\}$ denotes the trace of a matrix, and adj () denotes the adjugate matrix. Hence we get for the first derivative,

$$\frac{\partial \lambda_l}{\partial r_i} = \frac{\operatorname{tr}\left\{\operatorname{adj}\right\}\left(M\left(\lambda_i\right)\right)\left(\partial/\partial r_i\right)H^T H}{\operatorname{tr}\left\{\operatorname{adj}\right\}\left(M\left(\lambda_i\right)\right)} \tag{6.13}$$

6.4.2 *The first derivative of the eigenvectors of $H^T H$*

We have

$$M\left(\lambda_l\right) q_l = 0$$

Hence,

$$\frac{\partial M\left(\lambda_l\right)}{\partial r_i} q_l + M\left(\lambda_l\right)\frac{\partial q_l}{\partial r_i} = 0$$

or

$$M\left(\lambda_l\right)\frac{\partial q_l}{\partial r_i} = -\frac{\partial M\left(\lambda_l\right)}{\partial r_i} q_l \tag{6.14}$$

This equation cannot be solved directly because $M\left(\lambda_i\right)$ is singular. Therefore, we proceed as follows. Note that by Eq. (6.14) and by the symmetry of $M(\lambda_i)$,

$$q_l^T M(\lambda_l) \frac{\partial q_l}{\partial r_i} = -q_l^T \frac{\partial M(\lambda_l)}{\partial r_i} q_l \tag{6.15}$$

Also, since q_l is a unit vector,

$$q_l^T q_l = 1; \quad \text{hence,} \quad q_l^T \frac{\partial q_l}{\partial r_i} = 0 \tag{6.16}$$

Consider Eq. (6.14) after pre-multiplying by Q^T:

$$\begin{bmatrix} \lambda_l - \lambda_1 & & & & & \\ & \ddots & & & 0 & \\ & & \lambda_l - \lambda_{1-1} & & & \\ & & 0 & & & \\ 0 & & & & \ddots & \\ & & & & & \lambda_1 - \lambda_k \end{bmatrix} Q^T \frac{\partial q_l}{\partial r_i} = -Q^T \frac{\partial M(\lambda_l)}{\partial r_i} q_i \tag{6.17}$$

Some manipulation of Eqs. (6.15)–(6.17) yields

$$\frac{\partial q_i}{\partial r_i} = -Q \begin{bmatrix} \lambda_l - \lambda_1 & & & & & \\ & \ddots & & & 0 & \\ & & \lambda_l - \lambda_{1-1} & & & \\ & & 1 & & & \\ 0 & & & & \ddots & \\ & & & & & \lambda_1 - \lambda_k \end{bmatrix}^{-1} \times Q^T \left(\frac{\partial \lambda_l}{\partial r_i} I - \frac{\partial}{\partial r_i} H^T H \right) q_l \tag{6.18}$$

6.4.3 *An expression for adj (M)*

In order to compute the second derivatives of the eigenvalues, we will need an explicit expression for $\text{adj}(M(\lambda_l))$. Since $M(\lambda_l)$ is singular, it is meaningless to use $\text{adj}(M(\lambda_l)) = \det(M(\lambda_l))[M(\lambda_l)]^{-1}$, and we will have to use a limiting process. For all $\lambda \neq \lambda_1, \lambda_2, \ldots, \lambda_k$, we have

$$M^{-1}(\lambda) = Q(\lambda I - \Lambda)^{-1} Q^T, \quad \det M(\lambda) = \prod_{m=1}^{k} (\lambda - \lambda_m)$$

Hence,

$$\text{adj}(M(\lambda)) = Q\Psi Q^T$$

where Ψ is a diagonal matrix whose l-th diagonal entry is $\Pi_{m\neq l}$ $(\lambda - \lambda_m)$.

Now let λ approach λ and use the fact that adj $(M(\lambda))$ is a continuous function of λ. Clearly, all the entries of Ψ go to zero, except for the l-th one that goes to $\Pi_{m\neq 1}(\lambda_l - \lambda_m)$. We therefore get

$$\text{adj}(M(\lambda_l)) = \left[\prod_{m\neq l} (\lambda_l - \lambda_m)\right] q_l q_l^T \tag{6.19}$$

Since all the eigenvalues of $H^T H$ were assumed to be distinct, adj$(M(\lambda))$ is a matrix of rank 1. We also see from Eq. (6.19) that

$$\text{tr}[\text{adj}\{M(\lambda_l)\}] = \prod_{m\neq l} (\lambda_l - \lambda_m) \neq 0$$

i.e., the denominator in Eq. (6.13) is nonzero.

6.4.4 *The second derivatives of the eigenvalues*

Differentiation of Eq. (6.13) with respect to r yields

$$\frac{\partial^2 \lambda_l}{\partial r_i \partial r_j} = \frac{\text{tr}\left\{\left(\frac{\partial\, \text{adj}(M(\lambda_l))}{\partial r_j}\right)\left(\frac{\partial}{\partial r_i}\right) H^T H + \text{adj}(M(\lambda_l))\left(\frac{\partial^2}{\partial r_i \partial r_j}\right) H^T H\right\}}{[\text{tr}\{\text{adj}(M(\lambda_l))\}]}$$
$$- \frac{\text{tr}(\partial\, \text{adj}(M(\lambda_l))/\partial r_j) \cdot \text{tr}\{\text{adj}(M(\lambda_l))(\partial/\partial r_i) H^T H\}}{[\text{tr}\{\text{adj}(M(\lambda_l))\}]^2} \tag{6.20}$$

In order to use this formula, we only need an expression for the first derivative of $\mathrm{adj}(M(\lambda_l))$. Using Eq. (6.19), we get

$$\frac{\partial\, \mathrm{adj}\,(M(\lambda_l))}{\partial r_j} = \left[\frac{\partial}{\partial r_j}\prod_{m\neq l}(\lambda_i - \lambda_m)\right] q_l q_l^T + \left[\prod_{m\neq l}(\lambda_i - \lambda_m)\right]\left[\frac{\partial q_l}{\partial r_j} q_l^T + q_l \frac{\partial q_l^T}{\partial r_j}\right] \tag{6.21}$$

The right-hand side of Eq. (6.21) can be computed from the previous expressions. Hence, we have all the formulas needed to compute the second derivatives. We just mentioned that the derivatives of $H^T H$ are easy to compute because the entries are just the various components of r. We have

$$\frac{\partial}{\partial r_i} H^T H = \frac{\partial H^T}{\partial r_i} H + H^T \frac{\partial H}{\partial r_i} \tag{6.22}$$

$$\frac{\partial^2}{\partial r_i \partial r_j} H^T H = \frac{\partial H^T}{\partial r_i}\frac{\partial H}{\partial r_j} + \frac{\partial H^T}{\partial r_j}\frac{\partial H}{\partial r_i} \tag{6.23}$$

The matrices $\partial H/\partial r_i$ and $\partial H/\partial r_j$ are constant matrices, consisting of 1s and 0s only.

6.5 Conclusion

This chapter has presented a method for identifying a minimal-order linear system from input-output data. The crucial problem is identifying the minimal order. By identifying maximal sets of linearly independent rows and linearly independent columns of a finite dimensional Hankel matrix, both the minimal order and a basis for the representation are identified. A statistical inference technique, based on sequential tests on the ranks of submatrices of the Hankel matrix of sample correlations and on an approximate distribution for the rank test statistics has been derived. The approximate distribution converges to the true one as the number of observations becomes large. The proposed method seems to close a gap between the minimal realization problem and the problem of system identification from input-output measurement data.

Appendix A.6 The covariance of the normal vector $\tilde{r}$ appearing in the quadratic form

It was shown in Section 6.3 that in order to specify the test statistics distribution, the value of $\Sigma = \text{cov}(\tilde{r})$ is needed. We have

$$E\left\{\tilde{r}\tilde{r}^T\right\} = E\left\{\hat{r}\hat{r}^T\right\} - rr^T$$

Since the vector r is constructed from elements of the correlations R_i, $i = 1, 2, \ldots$, the value of rr^T can be immediately obtained in terms of cross products $R_i R_j, i, j = 1, 2, \ldots$. We next derive expressions for the second moments of the entries of $\hat{R}_t$ and $\hat{R}_k$, from which $E\left\{\hat{r}\hat{r}^T\right\}$ is immediately obtainable.

Let the sample covariances $\left\{\hat{R}_i\right\}$ be defined by

$$\hat{R}_i = \frac{1}{T-1}\sum_{t=l+1}^{T} y_t u_{t-l}^T \tag{A.6.1}$$

The (i, j)-th entry of $\hat{R}_l$ will be denoted $\hat{R}_{l,ij}$. The sample covariances thus defined are clearly unbiased, i.e.,

$$E\left\{\hat{R}_{l,ij}\right\} = \frac{1}{T-l}\sum_{t=l+1}^{T} E\left\{y_{t,i}u_{t-l,j}\right\} = R_{l,ij}$$

The second moments of the sample covariances are computed as follows.

$$\begin{aligned}
\hat{R}_{l,ij}\hat{R}_{k,gh} &= \left(\frac{1}{T-l}\sum_{t=l+1}^{T} y_{t,i}u_{t-i,j}\right)\left(\frac{1}{T-k}\sum_{s=k+l}^{T} y_{s,g}u_{s-k,h}\right) \\
&= \frac{1}{(T-k)(T-1)}\sum_{t=l+1}^{T}\sum_{s=k+1}^{T} y_{t,i}u_{t-l,j}y_{s,g}u_{s-k,h}
\end{aligned}$$

Since $\{u\}$ and $\{y)$ are jointly Gaussian, we have

$$\begin{aligned}
E\left\{\hat{R}_{t,ij}\hat{R}_{k,hg}\right\} &= \frac{1}{(T-k)(T-1)}\sum_{t=l+1}^{T}\sum_{s=k+1}^{T} \\
&\quad \times \left\{\begin{array}{l}
E\left\{y_{t,i}u_{t-l,j}\right\}E\left\{y_{s,g}u_{s-k,h}\right\} \\
\quad + E\left\{y_{t,i}y_{s,g}\right\}E\left\{u_{t-l,j}u_{s-k,h}\right\} \\
\quad + E\left\{y_{t,i}u_{s-k,h}\right\}E\left\{u_{t-l,j}y_{s,g}\right\}
\end{array}\right\}
\end{aligned}$$

Hence

$$\operatorname{cov}\left\{\hat{R}_{l,ij}, \hat{R}_{k,gh}\right\} = \frac{1}{(T-l)(T-k)} \sum_{t=l+1}^{T} \sum_{s=k+l}^{T} \times \left\{ \begin{array}{l} E\left\{y_{t-s,i} y_{0,g}\right\} E\left\{u_{t-l-s+k,j} u_{0,k}\right\} \\ \quad + E\left\{y_{t-s+k,i} u_{0,h}\right\} E\left\{u_{t-l+s,j} y_{0,g}\right\} \end{array} \right\} \tag{A.6.2}$$

Note that, since $\{u_i\}$ is uncorrelated, we have

$$E\left\{u_{t-l+s-k} u_{0,h}\right\} = 0, \quad t-l+s-k \neq 0 \tag{A.6.3}$$

In practice, the expectations appearing in Eq. (A.6.2), which are unknown, are approximated by

$$E\left\{y_{m,i} y_{0,j}\right\} \approx \frac{1}{T-m} \sum_{t=m+1}^{T} y_{t,i} y_{t-m,j} \tag{A.6.4a}$$

$$E\left\{y_{m}, u_{0,j}\right\} \approx \frac{1}{T-m} \sum_{t=m+l}^{T} y_{t,i} u_{t-m,j} \tag{A.6.4b}$$

$$E\left\{u_{0} u_{0,j}\right\} \approx \frac{1}{T} \sum_{t=l}^{T} u_{t,i} u_{t,j} \tag{A.6.4c}$$

Chapter 7

Minimal-Order Predictability of Future States

7.1 Introduction

The prediction of a vector-valued stationary sequence possessing a linear representation of finite order can be approached, in principle by first constructing such a representation from the process of second-order statistics and then matching a Kalman filter to it. The construction of a linear representation can be performed by spectral factorization (Anderson, 1969) or by stochastic realization (Faurre, 1969). Given the covariance function of the process in the time domain, both approaches would be indirect, as the first requires the transformation to the frequency domain, while the second requires the intermediate solution of a different problem, namely, the deterministic input-output realization problem. The construction of the Kalman filter further requires the solution of a Riccati equation. More direct predictor construction from the covariance function (Faurre, 1976; Son and Aderson, 1973) also requires the solution of Riccati equations.

Here, following an earlier observation (Baram, 1990), we propose a direct approach to the construction of linear predictors for stationary vector sequences from the covariance function. The approach, inspired by the coordinate-free realization concept (Akaike, 1974a; 1974b), is based on geometric principles. It suggests an explicit coordinate-dependent characterization of the class of all minimal-order linear predictors, in terms of linear transformations on the Hankel covariance matrix associated with the dynamical sequence at hand. It does not give rise to Riccati or Lyapunov equations.

The selection of particular transformation matrices defines specific predictor construction techniques.

7.2 Generalized formulation

Let the covariance function of a zero-mean, full-rank stationary sequence $y \in R^m$ be given by

$$R_k = E\{y_n y_{n-k}^T\} \tag{7.1}$$

where E denotes the expectation operation, and suppose that there exist a positive integer P and a set of scalars $a_1, \ldots, a_{P-1}, a_0 = 1$, such that for any $n \geq P$

$$\sum_{k=0}^{P-1} a_k R_{n-k} = 0 \tag{7.2}$$

It is well known that when y_n is generated by a finite-dimensional linear system driven by white noise, a relationship of the type Eq. (7.2) is satisfied.

The prediction problem is one of estimating the values of y_{n+k}, $k = 0, 1, \ldots,$ given the values of $y_{n-1}, y_{n-2}, \ldots$. Let us denote by $Y_n^- = (y_{n-1}^T, y_{n-2}^T, \ldots, y_0^T)^T$ and $Y_n^+ = (y_n^T, y_{n+1}^T, \ldots,)^T$ the past and the future vectors at time n. Let $y_{k,i}$ denote the i-th element of y_k and let $y_{k|n-1}$ denote the linear mean-square projection of y_k on Y_n^-. Let us further denote by $(Y_n^+|Y_n^-)$ the space generated by $\{y_{k,i|n-1,\ i=1,\ldots,m,k\geq n}\}$ by $Y_n^-(k) = (y_{n-1}^T, y_{n-2}^T, \ldots, y_{n-k}^T)^T$ and $Y_n^+(k) = (y_n^T, y_{n+1}^T, \ldots, y_{n+k-1}^T)^T$ the k-step past and future vectors at time n, and by $Y_{n|n-1}(k)$ the linear mean-square projection of $Y_n^+(k)$ on Y_n^-.

We have (see, e.g., (Anderson and Moore, 1979))

$$Y_{n|n-1}^+(k) = R(k,n)[E\{Y_n^- Y_n^{-T}\}]^{-1} Y_n^-$$

where

$$R(k,n) = E\{Y_n^+(k) Y_n^-\} = \begin{bmatrix} R_1 \; R_2 \cdots R_n \\ R_2 \; R_3 \cdots R_{n+1} \\ \vdots \\ R_k \; R_{k+1} \cdots R_{n+k-1} \end{bmatrix}$$

It follows from Eq. (7.2) that for any $k \geq P$ the rows beyond the first $P \cdot m$ rows of $R(k, n)$ are linearly dependent on the previous ones. The space $(Y_n^+ | Y_n^-)$ is then spanned by

$$Y_{n|n-1}^+(P) = R(P, n)[E\{Y_n^- Y_n^{-T}\}]^{-1} Y_n^-$$

It can also be seen from Eq. (7.2) that the columns of $R(k, n)$ beyond the first $P \cdot m$ are linearly dependent on the previous ones. It follows that the first maximal set of linearly independent rows of $R(P, n)$ is indexed as the set of such rows of the matrix

$$R = E\{Y_n^+(P)[Y_n^-(P)]^T\} = \begin{bmatrix} R_1 \; R_2 \cdots R_P \\ R_2 \; R_3 \cdots R_{P+1} \\ \vdots \\ R_P \; R_{P+1} \cdots R_{2P-1} \end{bmatrix}$$

The subsequent analysis will require some further notation. Let A and B be any two matrices such that A has no fewer rows than B. Let us denote by $r_B(A)$ the matrix whose rows are those rows of A, indexed as the first maximal set of linearly independent rows of B. Similarly, let us denote by $c_B(A)$ the matrix whose columns are those columns of A indexed as the first maximal set of linearly independent columns of B. We will find it convenient to write these row and column selection operations in terms of matrix products

$$r_B(A) \equiv r_B^A A$$

and

$$c_B(A) \equiv A C_B^{A^T}$$

where r_B^A and c_B^A are properly dimensioned, full rank matrices of zeros and ones. For instance, suppose that A has dimension 5×7 and that rows 1, 3 and 4 and columns 1, 2 and 5 of B form maximal independent sets. Then

$$r_B^A = \begin{bmatrix} 1\,0\,0\,0\,0 \\ 0\,0\,1\,0\,0 \\ 0\,0\,0\,1\,0 \end{bmatrix}$$

and

$$c_B^A = \begin{bmatrix} 1\,0\,0\,0\,0\,0\,0 \\ 0\,1\,0\,0\,0\,0\,0 \\ 0\,0\,0\,0\,1\,0\,0 \end{bmatrix}$$

Since it may be assumed that the matrices involved have compatible dimensions, we will use the somewhat abbreviated notation

$$r_B \equiv r_B^A, \quad c_B \equiv c_B^A$$

Let M and N be arbitrary nonsingular square matrices of dimension $m \cdot P$ and let

$$X_n = MY_{n|n-1}^+(P)$$

and

$$Z_n = NY_n^-(P)$$

Let us further define

$$x_n = r_{MRN^T} X_n$$

and

$$z_n^T = Z_n^T c_{MRN^T}$$

It can be seen that x_n is a vector of minimal dimension, which spans $(Y_n^+|Y_n^-)$ in the sense that each element of the latter space can be obtained by a linear combination of the elements of x_n. Clearly, x_{n+1} belongs to $(Y_{n+1}^+|Y_{n+1}^-)$, which is included in $(Y_n^+|Y_{n+1}^-)$. The latter may be decomposed as $(Y_n^+|Y_{n+1}^-) = (Y_n^+|Y_n^-) \otimes V_n$, where V_n is the space generated by

$$v_n = y_n - y_{n|n-1} \tag{7.3}$$

and $\otimes$ denotes the Cartesian product. Since V_n is orthogonal to $(Y_n^+|Y_n^-)$, it follows that there exist matrices A_n and B_n such that

$$x_{n+1} = A_n x_n + B_n v_n \tag{7.4}$$

Multiplying Eq. (7.4) on the right by z_n^T, taking expectation and noting that v_n is orthogonal to Y_n^-, hence, to z_n, we obtain

$$E\{x_{n+1}z_n^T\} = A_n E\{x_n z_n^T\} \tag{7.5}$$

where

$$\begin{aligned} E\{x_n z_n^T\} &= r_{MRN^T} M E\{Y_{n|n-1}(P)[Y_n^-(P)]^T\} N^T c_{MRN^T}^T \\ &= r_{MRN^T} MRN^T c_{MRN^T}^T \\ &= \bar{M}R\bar{N}^T \end{aligned} \tag{7.6}$$

where

$$\bar{M} = r_{MRN^T} M, \quad \bar{N} = c_{MRN^T} N$$

It can be seen that $E\{x_n z_n^T\}$ is a nonsingular matrix. We also have

$$\begin{aligned} E\{x_{n+1}z_n^T\} &= r_{MRN^T} M E\{Y_{n+1|n}(P)[Y_n^-(P)]^T\} N^T c_{MRN^T}^T \\ &= \bar{M}R^s\bar{N}^T \end{aligned} \tag{7.7}$$

where

$$R^s = E\{Y_{n+1}^+(P)[Y_n^-(P)]^T\} = \begin{bmatrix} R_2 & R_3 & \cdots R_{P+1} \\ R_3 & R_4 & \cdots R_{P+2} \\ \vdots & & \\ R_{P+1} & R_{P+2} & \cdots R_{2P} \end{bmatrix} \tag{7.8}$$

Substituting Eqs. (7.6) and (7.7) into Eq. (7.5), we obtain

$$A_n = A = \bar{M}R^s\bar{N}^T(\bar{M}R\bar{N}^T)^{-1}$$

It follows from the definition of x_n that there exists a matrix C_n, such that

$$y_n = C_n x_n + v_n \tag{7.9}$$

Multiplying Eq. (7.9) on the right by z_n^T and taking expectation, we obtain

$$C_n = E\{y_n z_n^T\}[E\{x_n z_n^T\}]^{-1} \tag{7.10}$$

where

$$E\{y_n z_n^T\} = (MRN^T)_1 c^T_{MRN^T}$$

where $(MRN^T)_1$ is the first block row of MRN^T. Hence

$$C_n = C = (MRN^T)_1 c^T_{MRN^T} (\bar{M}R\bar{N}^T)^{-1}$$

Multiplying Eq. (7.4) on the right by υ_n^T and taking expectation, noting that υ_n is orthogonal to x_n, we obtain

$$B_n = B = E\{x_{n+1}\upsilon_n^T\}[E\{\upsilon_n \upsilon_n^T\}]^{-1} \tag{7.11}$$

where

$$E\{x_{n+1}\upsilon_n^T\} = E\{x_{n+1}y_n^T\} - E\{x_{n+1}x_n^T\}C^T = \bar{M}\underline{R} - A\Pi C^T \tag{7.12}$$

and

$$E\{\upsilon_n \upsilon_n^T\} = E\{(y_n - Cx_n)y_n^T\} = R_0 - C\Pi C^T \tag{7.13}$$

where

$$\Pi = E\{x_n x_n^T\}$$

and

$$R = (R_1^T R_2^T \cdots R_P^T)^T \tag{7.14}$$

We next derive the term ΠC^T, which appears in both Eq. (7.12) and Eq. (7.13). Noting that

$$R_k = E\{y_{n+k}y_n^T\} = CA^k \Pi C^T$$

it can be seen that

$$\Theta \Pi C^T = \underline{R}$$

where $\underline{R}$ is defined by Eq. (7.14) and

$$\Theta = \begin{bmatrix} CA \\ CA^2 \\ \vdots \\ CA^P \end{bmatrix} \tag{7.15}$$

Since $Y_{n|n-1}(P) = \Theta x_n$ and since the covariance rank of x_n is the same as that of $Y_{n|n-1}(P)$, it follows that Θ is full rank. Hence,

$$\Pi C^T = (\Theta^T \Theta)^{-1} \Theta^T \underline{R}$$

which, substituted in Eqs. (7.12) and (7.13), completes the derivation of the matrix B, defined by Eq. (7.11). In summary, the generalized predictor is given by

$$\begin{aligned} x_{n+1} &= A x_n + B \upsilon_n, \;\; x_0 = 0 \\ y_{n+k|n-1} &= C A^k x_n \end{aligned} \tag{7.16}$$

where

$$A = \bar{M} R^s \bar{N}^T (\bar{M} R \bar{N}^T)^{-1} \tag{7.17}$$

$$C = (MRN^T)_1 c^T_{MRN^T} (\bar{M} R \bar{N}^T)^{-1} \tag{7.18}$$

$$B = [\bar{M} - A(\Theta^T \Theta)^{-1} \Theta^T] \underline{R} [R_0 - C(\Theta^T \Theta)^{-1} \Theta^T R] \tag{7.19}$$

and

$$\upsilon_n = y_n - C x_n$$

The class of all minimal-order linear predictors for the sequence y_n is now defined by Eq. (7.16). A particular member of the class is specified by selection of the nonsingular matrices M and N. Two particular selections are suggested in the following section. We note that

$$\begin{aligned} x_{n+1} &= A x_n + B \upsilon_n \\ y_n &= C x_n + \upsilon_n \end{aligned} \tag{7.20}$$

with A, B, C and υ_n as defined above, is a minimal realization of the sequence y_n. (As we have noted in previous sections, the minimality of required resources would seem to be a natural preference in the evolutionary design of biological mechanisms. From a scientific viewpoint, the establishment of a mathematical theory supporting such evolutionary preference would appear to be particularly rewarding.)

7.3 Specific constructions

7.3.1 *Original coordinates*

When the matrices M and N are taken to be identity matrices, the vectors $Y_n^-(P)$ and $Y_n^+(P)$ are left in their original coordinates. The minimal predictor for this choice is specified in the preceding section, with MRN^T and MR^sN^T replaced simply by R and R^s. The predictor is essentially defined by maximal sets of linearly independent rows and linearly independent columns of R. The following procedure selects a maximal set of such rows. Extension to the selection of linearly independent columns is immediate.

The procedure consists of checking the rank deficiency of matrices R_i, consisting of the first $i-1$ linearly independent rows of R and the rows following them. If R_i is rank-deficient, its last row is replaced by the next row of R and the rank check is repeated. If R_i is full-column rank, the next row of R is attached to R_i to form R_{i+1} and a rank check is performed. This procedure is continued until rank $R_i = M$ or until all the rows of R have been checked. The final R_i will contain a maximal set of linearly independent rows of R. This procedure can be extended to the construction of a linear predictor or a minimal realization from a sample covariance sequence, employing a proposed statistical rank test method (Baram and Porat, 1988).

7.3.2 *Canonical coordinates*

Construction in the original coordinates requires a sequence of rank checks on submatrices of the Hankel covariance matrix R. A computationally more attractive technique may be obtained by selecting the matrices M and N, so as to diagonalize the matrix MRN^T. Let M and N be the matrices of normalized eigenvalues of RR^T and R^TR, respectively, corresponding to nonzero eigenvalues. These may be obtained directly from the singular value decomposition of R (see, e.g., (Golub and Kahan, 1965)). It can be readily seen that

$$r_{MRN^T} = c_{MRN^T} = [I_P\ 0] \tag{7.21}$$

Denoting

$$S = \begin{bmatrix} s_1 & & & 0 \\ & s_2 & & \\ & & \ddots & \\ o & & & s_p \end{bmatrix}$$

where $s_1, \ldots, s_p$ are the squared nonzero singular values of R, the predictor is now specified by the matrices

$$A = [I_p \; 0] M R^s N^T \begin{bmatrix} I_P \\ 0 \end{bmatrix} S^{-1} \tag{7.22}$$

$$C = [s_1 \; 0 \ldots 0] S^{-1} \tag{7.23}$$

and B is defined by Eq. (7.19), with r_{MRM^T} defined by Eq. (7.21), Θ is defined by Eq. (7.15), and A and C are defined by Eqs. (7.22) and (7.23). We note that while the above procedure is computationally more effective than the one performed in the original coordinates, when the exact covariance function is known, it is not extendable by means of statistical inference to the case where only a sample covariance sequence is given. We also note that the above construction is different than canonical variates realization (Larimore, 1983), which produces an approximate representation for the process, even when the latter possesses a finite order realization. The present construction produces an exact realization of the form Eq. (7.20) with the parameters specified above.

7.4 Conclusion

We have proposed a direct approach to the construction of a minimal-order linear predictor for a stationary vector sequence from the covariance function. The class of all such predictors has been characterized in terms of linear transformations of the Hankel covariance matrix. Two specific construction schemes have been specified. While, in a biological context, such mathematical constructions can only be assumed to be based on natural wisdom, artificially minimal sensory feedback can be incorporated by technologically means in a neurological context, as presented in the following chapters.

PART II

Multi-Modal Linear Control Systems

Chapter 8

Distance Measures for Stochastic Models

8.1 Introduction

Measures of the difference between statistical models for an observed process have long been of interest. The Bhattacharyya distance (Bhattacharyya, 1943; Kailath, 1967), the invariant prior probability (Jeffreys, 1946), and the Kullback–Leibler divergence (Kullback and Leibler, 1951) are a few examples. However, as noted by Kullback (1959), these measures are not proper metrics (or distance measures) on a given set of models, as they fail to satisfy the triangle inequality. A modified version of Kullback and Leibler's information has been shown to be a proper distance measure, but its computation has been limited to the case where the true distribution of the observations is known (Baram, 1976). A class of rather abstract metrics has been considered (Zolotarev, 1977) for the purpose of approximate distributions. The difficulty associated with computing or estimating such real metrics is that the true distribution of the observation is intrinsic in their definition. Even if a certain dynamical model for the process may be assumed, the problem of finding its probability distribution is still a difficult one, and has only been solved for limited classes of processes, such as linear and Gaussian. However, while the models used for estimation and control purposes will often be linear and Gaussian, the actual process will be neither linear nor Gaussian. The question is whether a distance between models in a given set can be computed or estimated without restrictive assumptions on the true model for the observations, which is a measure of the given model's compatibility as an approximation of the actual model.

In this chapter, following an earlier study (Baram 1981), we define the distance between two stochastic models simply as the absolute difference between the corresponding values of some criterion function. When the latter is based on a certain realization of the observation process, the distance between the two given models may be computed directly from the data. When the criterion function is based on the distribution of the observations, the distance between two given models can be computed if the true distribution is known or may be estimated, under certain conditions, from the data. As special cases, we examine the likelihood, or information, criterion and a normalized prediction criterion. The distance criterion may, but need not, correspond to a parameter estimation criterion. This is demonstrated by the fact that the normalized prediction criterion does not necessarily attain its minimum at the true parameter, as does the likelihood criterion. While the computation of the information distance between two models within a given set of linear Gaussian models presents no difficulty, a meaningful estimate of the distance between a given model and the true, unknown model is not obtainable. In contrast, the prediction distance between a given model and the true one can be estimated from an ergodic sequence of observations. The distance can be quite useful in evaluating a given linear model as an approximation of the true model. The distance measures are shown to have an analytic value in studying properties such as model distinguishability and model identifiability, and to be useful tools for model selection purposes.

8.2 General: Distance between stochastic models

Let $z_1, \ldots, z_n$ be a sequence of observations having a finite dimensional distribution $F_*(Z^n)$ on R^k where $Z^n \equiv (z_1, \ldots, z_n)$. Let $S \in R^p$ be a parameter set, such that for each parameter $s \in S$ and n, there exists a model for Z^n. We shall consider scalar functions $h_n(s) = h(s, Z^n)$ and the expectations of $h_n(s)$ with respect to $F_*(Z^n)$, $g_n(s) \equiv E_* h_n(s)$.

For any n and $h_n(s)$, we may define a pseudo-metric on S by

$$\partial_n(s,t) = |h_n(s) - h_n(t)| \tag{8.1}$$

and

$$d_n(s,t) = |g_n(s) - g_n(t)| \tag{8.2}$$

∂_n and d_n are metrics on S if and only if $h_n(s) = h_n(t)$ and $g_n(s) = g_n(t)$, respectively, implying $s = t$. The functions $h_n(s)$ and $g_n(s)$ then define maps between metric spaces

$$h_n : (S, \delta) \to (S, \partial_n) \tag{8.3}$$

and

$$g_n : (S, \delta) \to (S, d_n) \tag{8.4}$$

where δ is the standard Euclidean metric on R^p.

In the sequel, we shall refer to "the true parameter" denoted by $*$. In the case of $h_n(s)$, the true parameter may be regarded as the data set, or, abstractly, as the phenomenon which generated the data. In the case of $g_n(s)$, the true parameter $*$ is a vector whose elements are the scalar parameters of the true distribution $F_*(Z^n)$. In both cases, the parameter set S and the functions $h_n(s)$, $g_n(s), \partial_n(s)$ and $d_n(s)$ will extend in a natural manner to the set $(S \cup *)$.

8.3 The information distance

Suppose that for each parameter $s \in S$ there exist finite dimensional distributions $F_s(Z^n)$, densities $f_s(Z^n)$ and conditional densities $f_s(z_n|Z^{n-1})$ for the observation sequence (z_n) on R^k. We denote by E_s expectation with respect to F_s. Let us consider as distance criteria the functions

$$h_n(s) = h_n^{(1)}(s) \equiv \frac{1}{n} \log f_s(Z^n) = \frac{1}{n} \sum_{i=1}^{n} \log f_s(z_i|Z^{i-1}) \tag{8.5}$$

and

$$g_n(s) = g_n^{(1)}(s) = \frac{1}{n} \sum_{i=1}^{n} E \log f_s(z_i|Z^{i-1}) \tag{8.6}$$

and define the corresponding pseudo-metrics

$$\partial_n^{(1)}(s,t) \equiv |h_n^{(1)}(s) - h_n^{(1)}(t)| \tag{8.7}$$

and

$$d_n^{(1)}(s,t) \equiv |g_n^{(1)}(s) - g_n^{(1)}(t)| \tag{8.8}$$

Suppose that the model for (z_n) corresponding to each parameter $s \in S$ is Gaussian. Then we have

$$h_n^{(1)}(s) = -\frac{1}{2n}\left\{\sum_{i=1}^{n} \log|\Sigma_i(s)| + \sum_{i=1}^{n}[z_i - \hat{z}_i(s)]^T \Sigma_i^{-1}(s)[z_i - \hat{z}_i(s)]\right\} \tag{8.9}$$

and

$$g_n^{(1)}(s) = -\frac{1}{2n}\left\{\sum_{i=1}^{n} \log|\Sigma_i(s)| + \sum_{i=1}^{n} \operatorname{tr} \Sigma_i^{-1}(s)\Gamma_i(s)\right\} \tag{8.10}$$

where

$$\begin{aligned}
\hat{z}_i(s) &= E_s(z_i|Z^{i-1}) \\
\Sigma_i(s) &= E_s\{[z_i - \hat{z}_i(s)]^T[z_i - \hat{z}_i(s)]\} \\
\Gamma_i(s) &= E_*\{[z_i - \hat{z}_i(s)]^T[z_i - \hat{z}_i(s)]\}
\end{aligned}$$

The computation of $\partial_n^{(1)}(s,t)$ when the models corresponding to set S are linear and Gaussian presents no difficulty and is not dependent upon knowledge of the true model. On the other hand, the computation of $d_n^{(1)}(s,t)$ is practically limited to the case where the true parameter is known and is linear and Gaussian. Now suppose that the true model is unknown, but (z_n) may be assumed to be ergodic. Then $h_n^{(1)}(s)$ and $\partial_n^{(1)}(s,t)$ are consistent estimates of $g_n^{(1)}(s)$ and $d_s^{(1)}(s)$, respectively. Note that no further assumptions, such as linearity or normality on the true parameter, need be made.

Next, suppose that it is desired to estimate the information distance between the model corresponding to the parameter $s \in S$ and the true model. Then, both the estimation of $\partial_n^{(1)}(*,t)$ and $d_n^{(1)}(*,s)$ require the estimation of $\Sigma_n(*)$, which means, in essence, identifying the true model first. The identification can only be performed, for all practical purposes, on a set of linear models, which will normally be the model set corresponding to S. The resulting estimated distance will not be very useful then, as it will be nothing more than the distance between two models given on S.

8.4 The prediction distance

Now consider the criterion functions

$$h_n(s) = \tilde{h}_n^{(2)} \equiv \frac{1}{n}\sum_{i=1}^{n}[z_i - \hat{z}_i(s)]^T \Sigma_i^{-1}(s)[z_i - \hat{z}_i(s)] \tag{8.11}$$

and

$$g_n(s) = \tilde{g}_n^{(2)}(s) = \frac{1}{n}\sum_{i=1}^{n} \operatorname{tr} \Sigma_i^{-1}(s)\Gamma_i(s) \tag{8.12}$$

$\tilde{h}_n^{(2)}(s)$ and $\tilde{g}_n^{(2)}(s)$ are measures of the error in predicting the observations z_i by a predictor based on the parameter s. The role of the normalizing matrix $\Sigma_n^{-1}(s)$ is evident from the fact that

$$\tilde{g}_n^{(2)}(*) = k \tag{8.13}$$

Let us further normalize

$$h_n^{(2)} = \frac{1}{k}\tilde{h}_n^{(2)}(s) \tag{8.14}$$

and

$$g_n^{(2)}(s) = \frac{1}{k}\tilde{g}_n^{(2)}(s) \tag{8.15}$$

The distance measures defined on $h_n^{(2)}(s)$ and $g_n^{(2)}(s)$ are

$$\partial_n^{(2)}(s,t) = |h_n^{(2)}(s) - h_n^{(2)}(t)| \tag{8.16}$$

and

$$d_n^{(2)}(s,t) = |g_n^{(2)}(s) - g_n^{(2)}(t)| \tag{8.17}$$

In particular, we have

$$d_n^{(2)}(*,s) = |1 - g_n^{(2)}(s)| \tag{8.18}$$

Note that while $\partial^{(1)}$ and $d^{(1)}$ were defined on the probability density of the observation sequence, $\partial^{(2)}$ and $d^{(2)}$ take into account its second moment only.

As in the case of the information distance, while the computation of $d_n^{(2)}(s,t)$ is practically limited by the linear-Gaussian assumption on the true parameter, $\partial_n^{(2)}(s,t)$ is a consistent estimate of $d_n^{(2)}(s,t)$ under the ergodicity assumption on (z_n). However, in contrast to the information distance, the prediction distance $d_n^{(2)}(*,s)$ between parameters in the set S and the true parameter can be estimated quite easily. Indeed

$$\hat{d}_n^{(2)}(*,s) \equiv |1 - h_n^{(2)}(s)| \tag{8.19}$$

is a consistent estimate of $d_n^{(2)}(*,s)$. Note that the validity of $\hat{d}_n^{(2)}$ is independent of linearity or normality assumptions on the true model, and its computation does not require model identification, as does the estimation of the information distance.

It should be noted that while the minimization of $h_h^{(1)}(s)$ is a proper parameter estimation criterion (it is the maximum likelihood criterion), the minimization of $h_h^{(2)}(s)$ is not. In particular, in the ergodic case

$$g_n^{(2)}(s) = \lim_{n\to\infty} h_n^{(2)}(s) \tag{8.20}$$

does not necessarily attain its minimum at the true parameter. This is demonstrated by the following example:

Let the model for z_n be given by

$$M_s : z_n = a + v_n \tag{8.21}$$

where a is a known constant scalar and v_n is a zero-mean white noise sequence with covariance r. The least-squares estimate (one-step prediction) of z_n is

$$\hat{z}_n(s) = a \tag{8.22}$$

and the estimation error covariance is

$$\sigma(s) = r \tag{8.23}$$

Suppose that the true model for z_n is

$$M_* : z_n = a + \omega_n \tag{8.24}$$

where ω_n is a zero-mean white noise sequence with covariance q such that $q < r$. Then

$$\gamma(s) \equiv E_*\{[z_n - \hat{z}_n(s)]^2\}q \tag{8.25}$$

Hence

$$g^{(2)}(s) = q/r < 1 = g^{(2)}(*) \tag{8.26}$$

which shows that $g^{(2)}$ is not minimal at the true parameter.

Note that if instead of $\tilde{h}_n^{(2)}(s)$, one takes

$$v_n(s) = \frac{1}{n}\sum_{i=1}^{n}[z_i - \hat{z}_i(s)]^T[z_i - \hat{z}_i(s)] \tag{8.27}$$

then

$$\min_s v_n(s) \tag{8.28}$$

is a proper parameter estimation criterion – the prediction error criterion (Ljung, 1975). Ljung has shown that the identification technique, based on Eq. (8.28), converges to a subset of the parameter set, which minimizes

$$l_n(*, s) \equiv E_*\{|\hat{y}_n(*) - \hat{y}_n(s)|^2\} \tag{8.29}$$

Note that although it does not have the exact form of our distance definition Eq. (8.1), the measure

$$l_n(s, t) = E_*\{|y_n(s) - y_n(t)|^2\} \tag{8.30}$$

is a proper distance measure on the parameter set. If one wishes to estimate this distance between a parameter s and the true parameter, then the same difficulties arise as with the information distance $d_n^{(1)}(*, s)$. The resulting distance, whose estimation would require excessive computation, will only be a distance between two parameters in S. Yet, the distance measures $l_n(*, s)$ and $d_n^{(1)}(*, s)$

have analytic value, as demonstrated by the convergence results in (Ljung, 1975) and (Baram, 1976).

In addition to the ability to estimate the distance $d_n^{(2)}(*, s)$ from ergodic observations, this distance measure has the advantage that it may be compared to the known limit value $g_n^{(2)}(*) = 1$ of $h_n^{(2)}(*)$. This, in turn, may be used to evaluate the compatibility of the model corresponding to s as an approximation of the true model. This property is discussed further in Section 8.6.

8.5 Parameter distinguishability and identifiability

An essential question in modeling considerations is whether different models in a given set can be distinguished, using a finite or an infinite data sequence. The difference between the models will normally be defined by the difference between the corresponding values of some criterion function, i.e., by the distance measure defined on that criterion.

For a given n and for a given criterion $h_n(s)$, a parameter s is said to be locally distinguishable if

$$\partial_n(s, s + ds) > 0 \tag{8.31}$$

for an infinitesimal change ds in the parameter value. Suppose that the criterion function $h_n(s)$ is continuously differentiable. Then

$$[\partial_n(s, s + ds)]^2 = [dh_n(s)]^2 = ds^T \left(\frac{\partial h_n(s)^T}{\partial s} \frac{\partial h_n(s)}{\partial s} \right) ds \tag{8.32}$$

It follows that a parameter $s \in S$ is locally distinguishable if and only if

$$P_n(s) = \frac{\partial h_n(s)^T}{\partial s} \frac{\partial h_n(s)}{\partial s} > 0 \tag{8.33}$$

For a given number of observations n, we shall say that a parameter $s \in S$ is globally distinguishable if, for any $s' \in S$, $s' \neq s$,

$$\partial_n(s, s') > 0$$

It follows that a pseudo-metric ∂_n is a metric on S if and only if each parameter in S is globally distinguishable.

It was argued in Section 8.1 that the distance criterion need not coincide with a parameter estimation criterion. When it does, the notions of local and global distinguishability are closely related to those of local and global identifiability used in the system identification literature, e.g., (Glover and Willems, 1974). For instance, it follows from a well-known equality (Kendall and Stuart, 1973, p. 9) that in the stationary case, when

$$h_n(s) = h_n^{(1)}(s) \tag{8.34}$$

then

$$E_s P_n(s) = E_s H_n(s) \tag{8.35}$$

when $H_n(s)$ is the Hessian, given by

$$(H_n)_{i,j}(s) = -\frac{\partial^2 h_n^{(1)}(s)}{\partial s_i \partial s_j} \tag{8.36}$$

and s_i and s_j are components of the parameter vector s. The Hessian plays a key role in the identification of Gaussian models. Tse (1973) has shown for stationary Gaussian models that if the expected value of $H(s)$, evaluated at the true parameter s^0, is positive definite, then s^0 is locally identifiable. Goodrich and Caines (1979) have shown that at the true parameter s^0, the sequence of Hessians $H_n(s^0)$ converges to a matrix $M(s^0)$, and that s^0 is locally identifiable if and only if $M(s^0) > 0$. It follows that s^0 is locally identifiable if and only if it is asymptotically locally distinguishable.

Global distinguishability is similarly related to global identifiability. It can be shown that each parameter in the set under consideration is globally distinguishable, then under certain differentiability conditions on $g_n(s)$, the estimates will converge to the parameter in the set which is closest to the true one in the distance measure associated with the estimation criterion (Caines, 1976; 1988). Ljung (1974b) has shown that the prediction error technique will converge to a subset of the parameter set, for which the distance in Eq. (8.30) is minimal. Global distinguishability will then imply that this subset contains but one parameter. Similarly, the maximum likelihood technique will converge, under global distinguishability in the distance $d_n^{(1)}$ to the parameter, which minimizes the distance to the true parameter. The relationship between consistency of parameter estimates and information measures has been pointed out by several authors, e.g., (Liporace, 1971; Akaike, 1972: Rissanen and Ljung, 1975: Hawkes and Moore, 1976).

8.6 Model selection

We have noted above that when the distance criterion coincides with a parameter estimation criterion, the estimation technique will, under a certain condition, converge to the parameter, which is closest to the true one in the distance measure. This suggests that the model selection problem may be approached directly by computing and minimizing the distance to the true model, rather than minimizing the estimation criterion. The minimizing distance can then be used to evaluate the compatibility of the selected model as an approximation of the true one. An examination of the distance measure considered in this chapter shows that this is not possible for the information distance, as the value of the likelihood function at the true parameter can only be estimated (or approximated) within the model set, which makes the proposed approach useless, for this criterion. On the other hand, the limit value of the prediction criterion $g_n^{(2)}$ at the true parameter is known, and, for an ergodic observation sequence, the prediction distance d_n^2 between a parameter in S and the true parameter is estimated by Eq. (8.19).

A model selection criterion that minimizes the estimated distance to the true parameter is then

$$\min_{s\in S} \hat{d}_n^{(2)}(*, s) = \min_{s\in S} |1 - h_n^{(2)}(s)| \tag{8.37}$$

When should the selected model be considered compatible? This may be decided by considering the size of the distance in Eq. (8.37) in absolute terms (e.g., 0.01 is small and 100 is large, relative to 1.0, the criterion function value at the true parameter) or by comparing it with distances within the model set. For instance, if $\bar{s}$ is the selected parameter, which satisfies Eq. (8.37), and if t is another similar parameter in the set S, then the selected model may be considered compatible if

$$\hat{d}_n^{(2)}(*, \bar{s}) = \min_{s \in S} \hat{d}_n^{(2)}(*, s) \leq \hat{d}_n^2(t, \bar{s}) = \partial_n^{(2)}(t, \bar{s}) \tag{8.38}$$

On the other hand, the selected model may be considered incompatible if the minimizing distance is of the same order of magnitude or greater than distances between parameters within the set S, which gives rise to "very different" models, or when

$$d_n^{(2)}(*, \bar{s}) \geq \max_{t \in S} \partial_n^2(t, \bar{s}) \tag{8.39}$$

Of course, more detailed considerations may apply to more specific problems.

8.7 Reduced-order modeling

Suppose that a linear and Gaussian model $M(s)$ for an observed process is given (whether it is correct or not is immaterial here), but its order is too high for implementation processes. A model of a lower order is desired. If the space of all linear-Gaussian models of a desired order is given by

$$M = (M(s), s \in S)$$

the selected model will be the projection in some metric of $M(s)$ onto the set M. Given a distance criterion $g_n(s)$, the selected model is the one that satisfies the criterion

$$\min_{s \in S} d_n(\bar{s}, s) \tag{8.40}$$

where d_n is the distance measure defined on $g_n(s)$. This is similar to the previous distance minimization problem, with $\bar{s}$ replacing the true parameter $*$, and $F_0(z^n)$, the observations distribution under $M(s)$, replacing the data. Note that either $d_n^{(1)}$ or $d_n^{(2)}$ may be used here, and the minimizing distance may be used to evaluate the compatibility of the selected model, as discussed above.

8.8 Selecting a typical model from a set

Supposed that it is desired to find a single model which is, in some sense, typical, or representative, of a given model set $M = (M(s), s \in S)$. We propose the following approach. Suppose that one of the parameters of the set S is correct, but is unknown. The selected model should be the one for which, given an appropriate distance criterion, its maximum possible distance from the true parameter is minimal. Let $g_{t,n}(s) = E_t h_n(s)$ for some criterion $h_n(s)$. (Note that t plays the role of the true parameter here.) Let $d_{t,n}(t, s)$ denote the distance measure defined on $g_{t,n}(s)$. Then, by the proposed approach, the selected model is the one that satisfies

$$\min_{s} \max_{s} d_{t,n}(t, s) \tag{8.41}$$

In particular, we have in the linear-Gaussian case

$$d^{(1)}_{t,n}(t, s) = |\log |\Sigma_n(s)| - \log |\Sigma_n(t)| + \operatorname{tr} \Sigma_n^{-1}(s)\Gamma_{t,n}(s) - k| \tag{8.42}$$

and

$$d^{(2)}_{t,n}(t, s) = |1 - \operatorname{tr} \Sigma_n^{-1}(s)\Gamma_{t,n}(s)| \tag{8.43}$$

where

$$\Gamma_{t,n}(s) = E_t\{[z_n - \hat{z}_n(s)]^T [z_n - \hat{z}_n(s)]\} \tag{8.44}$$

Note that although the definition of $d^1_{t,n}(t, s)$ does not require the linear assumption, and the definition of $d^2_{t,n}(t, s)$ does not require the linear-Gaussian assumption, the computation of the covariance matrices involved is, in practice, limited to the linear-Gaussian case.

8.9 Conclusion

In this chapter, we have presented a criterion-based generalized definition of distance measures for model selection. We have examined in detail the measures induced by two criteria, namely, the information, or likelihood criterion and the prediction criterion. In particular, the prediction distance has been proposed as a useful working tool for model selection. This simple measure has the unique property that the distance between a given model and the actual, unknown model can, under certain conditions, be estimated. The distance measure approach provides a framework for treating both analysis and design problems in model selection, facilitating highly intuitive geometric interpretation.

Chapter 9

Convergence on Finite Parameter Sets

9.1 Introduction

Before analyzing the selection of a linear control model from a finite set of such models, as done in the following chapters, we address the selection from a set of a more general models. The underlying process involves convergence of maximal likelihood and Bayesian parameter estimates over general classes of observation sequences (Baram, 1976, Chapter 4). The convergence of the estimates follows from the convergence of the likelihood ratios over the parameter set. Consistency conditions are derived in terms of the information in the observations. The case where the true parameter is not necessarily a member of the parameter set is also considered. Rates of convergence in the mean for the ML and the MAP procedures are derived.

9.2 Convergence of parameter estimates

Let (z_n) be a stochastic process on a probability space (Ω, U, P_K) and let $S = K \equiv \{0, \ldots, p\}$ be a parameter set such that $\{P_j; j \in K\}$ is a family of probability measures on (Ω, U). Let (U_n) be an increasing sequence of σ-subalgebras of U generated by (Z^n) and let $P_{j,n}$ be the restriction of P_j to U_n for each $j \in K$. Consider the following condition:

(c9.1) For some $k \in K$ and for each $j \in K$; $j \neq k$

$$\lim_{n \to \infty} \mathrm{h}_k^j(Z^n) = 0 \text{ a.e.} \tag{9.1}$$

where h is the map defined by Eq. (8.3). We next show that the convergence a.e. of the ML and the MAP parameter estimation procedures and the convergence a.e. and in m.s. of the LS procedure follow from condition (c9.1). Of course, the major difficulty in proving convergence of the parameter estimates is to verify condition (c9.1). In the sequel, we present conditions for general classes of observation sequences under which condition (c9.1) is satisfied when k is the true parameter and extend the results to the case where the true parameter is not necessarily a member of the parameter set. The latter case is treated specifically in a subsequent chapter where the following four theorems will prove very useful. Proofs of these theorems are provided in the closing Appendix of this book.

Theorem 9.1. *Suppose that condition* (c9.1) *is satisfied, then ML estimates on K converge a.e. to k as $n \to \infty$.* Δ

Theorem 9.2. *Under condition* (c9.1), *MAP estimates on K converge a.e. to k.* Δ

Theorem 9.3. *Suppose that a parameter vector s is assumed to belong to a finite set $s_j \in R^m$; $j \in K$ in the calculation of the estimates (but is not necessarily a member of the set). Suppose further that for some $k \in K$, condition* (c9.1) *is satisfied. Then LS estimates of s on R^m converge a.e. to s_k.* Δ

Theorem 9.4. *For the situation given in Theorem* 9.3, *LS estimates converge to s_k in the mean square.* Δ

9.3 Consistency of the estimates

Following Chapter 2 (Eq. (2.25)), we defined for each pair $k, j \in K$

$$I_n(k;j) \equiv E_k^{U_{n-1}} \log \mathrm{h}_j^k(z_n|Z^{n-1})$$

Let us also define

$$J_n(k;j) \equiv \log \mathrm{h}_j^k(z_n|Z^{n-1}) - I_n(k;j)$$

$J_n(k;j)$ is the error in the incremental information $I_n(k;j)$ or the information residual.

Denote

$$Y_n(k;j) \equiv \sum_{m=1}^{n} I_m(k;j)$$

and

$$V_n(k;j) \equiv \sum_{m=1}^{n} J_m(k;j)$$

Suppose that $* \in K$, i.e., that the true parameter is a member of the parameter set, and consider the following conditions:

(c9.2) For some $k \in K$ and for each $j \in K;\ j \neq k$

$$\limsup \mathrm{V}_n(k;j) > -\infty \text{ a.e.}$$

(c9.3) For some $k \in K$ and for each $j \in K;\ j \neq k$

$$\lim_{n\to\infty} Y_n(k;j) = \infty \text{ a.e.}$$

Lemma 9.1. *Suppose that conditions (c9.2) and (c9.3) hold for $k = *$. Then for each $j \in k;\ j \neq *$ we have*

$$\lim_{n\to\infty} \mathrm{h}_*^j(z^n) = 0 \text{ a.e.} \qquad \Delta$$

Theorem 9.5. *Suppose that some $k \in K$ is the true parameter. Then, under conditions (c9.2) and (c9.3), ML and MAP estimates are consistent a.e. and LS estimates are consistent a.e. and in the mean square.* Δ

Consider the following condition

(c9.4) For some $k \in K$ and for each $j \in K;\ j \neq k$, there exists some $\varepsilon_j > 0$ and a subsequence (n_{i_j}) of n, such that

$$I_{n_{i_j}}(k;j) \geq \varepsilon_j \text{ a.e. for all } n_{i_j}$$

Theorem 9.6. *Suppose that some $k \in K$ is the true parameter. Then, under conditions (c9.2) and (c9.4), ML and MAP estimates are consistent a.e. and in the mean square.* Δ

Lemma 9.2. *For any pair $j, k \in K$ let $(J_n(k;j))$ be an ergodic sequence. Then*

$$\limsup V_n(k;j) = \infty \text{ a.e.} \qquad \Delta$$

Lemma 9.3. *Suppose that for any $j, k \in K$, we have for any positive scalar a*

$$E_*\{V_{n_a}(k;j) - a\} < \infty \tag{9.2}$$

where

$$n_a = \inf\{n : V_n(k;j) > a\}$$

then, for each $\omega \in \Omega$, either

$$\lim_{n\to\infty} V_n(k;j,\omega) \text{ exists and is finite}$$

or

$$\limsup V_n(k;j,\omega) = \infty \qquad \Delta$$

9.4 Convergence in the absence of the true parameter

Consider the probability and parameter spaces given in Section 9.1. While the absolute continuity of the restrictions $P_{1,n}$ and $P_{2,n}$ of two measures P_1 and P_2 to the σ-subalgebra U_n of U is possible to verify in practical situations (it follows, e.g., from the absolute continuity of the corresponding conditional densities $\mathrm{f}_{1,n}(z_n|Z^{n-1})$ and $\mathrm{f}_{2,n}(z_n|Z^{n-1})$ for each n), the absolute continuity of P_1 and P_2 does not follow and is, in general, more difficult to verify. The following results are nevertheless interesting from a theoretical viewpoint.

Theorem 9.7. *Let conditions (c9.2) and (c9.3) hold for some parameter $k \in K$. Furthermore, suppose that the true measure P_* is absolutely continuous with respect to the measure P_k. Then, for each $j \in k$; $j \neq k$, we have*

$$\lim_{n\to\infty} \mathrm{h}_k^j(z^n) = 0$$

and, consequently, the parameter estimates will converge to the parameter k in the senses specified in Theorems 9.1 through 9.4. Δ

In the following section, we treat the case where the true parameter is not necessarily a member of the parameter set for a model of particular interest, namely, linear dynamical system. We shall not, however, investigate the absolute continuity of the probability measures P_* and P_k, but rather use simpler arguments, enabled by the particular problem under consideration.

Suppose that for any parameter $k \in K$, the restriction $P_{j,n}$ of the measure P_j; $j \in K$, $j \neq k$ is absolutely continuous with respect to the restriction $P_{k,n}$ of the measure P_k. An interesting observation is presented by the following theorem.

Theorem 9.8. *Suppose that condition* (c9.1) *holds for the parameter $k \in K$. Then, the measures P_j, $j \in K$; $j \neq k$ are singular with respect to the measure P_*.* Δ

9.5 L_1 convergence

The L_1 convergence of the likelihood and *a posteriori* probability ratios follows directly if condition (c9.1) holds. We show that under a certain condition on the information in the observation, the convergence rates are bounded by exponentials of the number of samples. The true parameter is not assumed to belong to the parameter set. These results provide performance measures for the ML and the Map estimation methods. In the following chapters, we show that bounds of the L_1 convergence rates can be computed in common situations for linear systems.

Theorem 9.9. *Suppose that condition* (c9.1) *holds for some $k \in K$. Then for each $j \in K$; $j \neq k$, we have*

$$\lim_{n\to\infty} E_* \mathrm{h}_j^k(z^n) = \infty$$

and

$$\lim_{n\to\infty} E_* \frac{\mathrm{f}^b(k|z^n)}{\mathrm{f}^b(j|z^n)} = \infty \qquad \Delta$$

Now consider the following condition.

(c9.5) There exists a parameter $k \in K$ such that for each $j \in K$; $j \neq k$, there exists a positive scalar α_j such that

$$\bar{I}_n(k;j) \geq \alpha_j \text{ for all } n \geq N_j \tag{9.3}$$

Theorem 9.10. *Under condition* (c9.5), *there exists some positive integer N such that for each $j \in K$; $j \neq k$, the sequences* $(\mathrm{h}_j^k(z^n))$ *and* $\left(\frac{\mathrm{f}^b(k|z^n)}{\mathrm{f}^b(j|z^n)}\right)$ *diverge in L_1 at rates no slower than exponential for all $n \geq N$.* Δ

At instant n the ML estimation method will select the parameter k if

$$\frac{\mathrm{f}_k(z^n)}{\mathrm{f}_j(z^n)} \geq 1 \text{ for all } j \in K;\ j \neq k \tag{9.4}$$

The MAP method will select k if

$$\frac{\mathrm{f}^b(k|Z^N)}{\mathrm{f}^b(j|Z^N)} \geq 1 \text{ for all } j \in K;\ j \neq k \tag{9.5}$$

9.6 Conclusion

Our consideration of model selection from a set, which may or may not include the correct model, was based on the statistical nature of the selection criterion, rather than the mathematical nature of the models. For instance, the L_1 convergence bounds established in Theorem 9.10 provide a qualitative measure of performance for the ML and the MAP estimates in terms of rates at which Eqs. (9.4) and (9.5) are attained in the mean. Of course, the bounds cannot be computed unless the true measure is known. Yet, if the true parameter can be assumed to belong to a finite set, then bounds can be computed over the set. This will be demonstrated in the following chapters, where we consider a specific type of model, namely, linear systems.

Chapter 10

Linear Model Selection from a Finite Set

10.1 Introduction

The formal information-theoretic implementation of the multi-modal analysis presented in the previous chapter, and dating back to the author's earlier work (Baram, 1976, Chapter 4), has further materialized as a multi-modal set analysis of linear systems (Baram 1976, Chapter 5; Baram and Sandell, 1978) presented in this chapter. Here, we restrict our attention to linear systems driven by white Gaussian inputs having time-invariant statistics. We make the assumption that the system has attained steady state, i.e., that all signals of interest are stationary. We first study the convergence of identification procedures. The convergence conditions are obtained in terms of the second-order statistics associated with the models in the model set. If the true model is included in the set, it will be identified under a verifiable uniqueness condition. If the true model is not included in the model set, then the identification procedures converge to a model in the model set, which is closest to the true model in the information metric sense introduced in Chapter 2 and in the sense of the second-order statistics associated with the model. Then we treat the L_1 convergence of the likelihood ratios and the ratios of the *a posteriori* probabilities. We show that under a simple uniqueness condition, the sequences of likelihood and *a posteriori* probability ratios are bounded in L_1 by simple exponentials. If the true system belongs to the given model set, then the bounds can be easily computed using the *a priori* data. The L_1 convergence results provide performance measures for the ML and the MAP identification methods. Finally, the analysis is extended to other modeling systems. Methods are suggested for selecting a reduced-order model to represent a high-order system and for selecting a representative model from a set from which the true system, or an appropriate model of it, are known to take their values.

The convergence of the identification procedures is proved by direct application of the ergodic theorem (Chapter 2, Theorem 2.5). The present chapter then depends only on the results of Chapter 2 and Section 9.1, while the more advanced probabilistic arguments in Chapter 9 are omitted (note that since we consider here a very specific class of observation sequences, we are, in fact, able to (Baram, 1981) treat a more interesting class of problems than that considered in Section 9.2, as the true parameter is not assumed to belong to the parameter set).

10.2 Linear systems

In the sequel, we consider linear systems driven by white Gaussian inputs noise. After deriving the relevant statistics associated with the model set under consideration, we obtain expressions for the information in the observations and for the information distance measure on the parameter set. Then we show that maximum likelihood and related identification procedures on finite sets of models converge to the model in the model set, which is closest to the actual system in the information distance measure. Finally, we establish criteria for the selection of a linear model from a given model set for certain generic classes of practical problems. In the context of this book, these would be, essentially, human movement mechanisms subject to neurological impairment.

As noted in Chapter 2, * marks a true model. Consider the system

$$\begin{aligned} x_{n+1} &= F_* x_n + G_* w_n \\ z_n &= H_* x_n + v_n \end{aligned} \tag{10.1}$$

initialized at $n = n_0$, with

$$E_*\{x_{n_0}\} = 0; \quad E_*\{x_{n_0} x_{n_0}^T\} = X_*$$

where (w_n) and (v_n) are zero-mean, white Gaussian sequences mutually uncorrelated and uncorrelated with x_{n_0}, with

$$E_*\{w_n w_n^T\} = Q_*; \quad E_*\{v_n v_n^T\} = R_*$$

The model set for Eq. (10.1) is represented by

$$M \equiv \{(F_s, G_s, H_s, Q_s, R_s);\ s \in \zeta\} \tag{10.2}$$

Let

$$T \equiv (* \cup \zeta)$$

(As in Chapter 9, the restriction to a finite set is done for the analysis of convergence and consistency. In Section 10.5, we consider other modeling problems and there the model set is allowed to be infinite. Also note that the results of the present chapter can easily be extended to the case where the system (Eq. (10.1)) is driven by an additional deterministic input sequence.)

Let

$$\hat{z} \equiv E_s\{z_n | Z^{n-1}\}; \quad s \in T$$

denote the one-step least-squares prediction of z_n, given the past observations Z^{n-1}, assuming that s is the true parameter (the predictor corresponding to s). Let

$$\Sigma_{s,n} = \Sigma_s(n, n_0) = E_s\{(z_n - \hat{z}_{s,n})(z_n - \hat{z}_{s,n})^T\} \tag{10.3}$$

and

$$\Gamma^t_{s,n} \equiv \Gamma^t_s(n, n_0) \equiv E_t\{(z_n - \hat{z}_{s,n})(z_n - \hat{z}_{s,n})^T\} \tag{10.4}$$

denote the prediction error covariance matrices according to the respective measures (for each $s \in T, z_{s,n}$ and $\Sigma_{s,n}$ are computed, in essence, by a Kalman filter corresponding to s). Denote

$$\Sigma_s \equiv \lim_{n_0 \to -\infty} \Sigma_s(n, n_0) \tag{10.5}$$

We shall use the following condition:

(c10.1) For each $s \in T, \Sigma_s$ exists and has a finite positive definite value.

Condition (c10.1) is satisfied if each model corresponding to $s \in T$ is detectable and controllable. For each $s \in T$, Σ_s is obtained by running to steady-state a Riccati equation, or, equivalently, a Kalman filter corresponding to $(F_s, G_s, H_s, Q_s, R_s)$.

Also denote

$$\Gamma_s^t \equiv \lim_{n_0 \to -\infty} \Gamma_s^t(n, n_0) \tag{10.6}$$

Provided that the limit exists, Γ_s^t is obtained by the following procedure. First, assuming that $n_0 = -\infty$; take $\Sigma_{s,n} = \Sigma_s$ for each $s \in T$ and for all $n \geq n_1$, where n_1 is any fixed integer. Assuming that t is the correct parameter, the dynamic equation generating simultaneously the state x and its one-step prediction by the Kalman filter corresponding to $s, \hat{x}_{s,n}$, is

$$\begin{bmatrix} x_{t,n+1} \\ \hat{x}_{s,n+1} \end{bmatrix} = \begin{bmatrix} F_t & 0 \\ F_s K_s H_t & F_s(I - K_s H_s) \end{bmatrix} \begin{bmatrix} x_{t,n} \\ \hat{x}_{s,n} \end{bmatrix} + \begin{bmatrix} G_t & 0 \\ 0 & F_s K_s \end{bmatrix} \begin{bmatrix} w_n \\ v_n \end{bmatrix}$$

where K_s is the Kalman gain corresponding to s, that is

$$K_s = \Sigma_s H_s^T (H_s \Sigma_s H_s^T + R_s)^{-1}$$

Let

$$F_s^t \equiv \begin{bmatrix} F_t & 0 \\ F_s K_s H_t & F_s(I - K_s H_s) \end{bmatrix}; \quad G_s^t \equiv \begin{bmatrix} G_t & 0 \\ 0 & F_s K_s \end{bmatrix}$$

and

$$Q^t \equiv \begin{bmatrix} Q_t & 0 \\ 0 & R_t \end{bmatrix}; \quad H_s^t \equiv [H_t - H_s]$$

Then the matrix

$$\Psi_{s,n}^t \equiv E\left\{ \begin{bmatrix} x_{t,n+1} \\ \hat{x}_{s,n+1} \end{bmatrix} [x_{t,n+1} \; \hat{x}_{t,n+1}] \right\} \tag{10.7}$$

is generated by the Lyapunov equation

$$\Psi^t_{s,n+1} = F^t_s \Psi^t_{s,n} F^{t^T}_s + G^t_s Q^t G^{t^T}_s \tag{10.8}$$

initiated at n_1 by any initial value. We can write

$$\Psi^t_{s.n} = \Psi^t_s(n, n_1)$$

Then let

$$\Psi^t_s = \lim_{n_1 \to -\infty} \Psi^t_s(n, n_1) \tag{10.9}$$

Finally

$$\Gamma^t_s = H^t_s \Psi^t_s H^{t^T}_s + R_t \tag{10.10}$$

It is well known that the limit Eq. (10.9) of Eq. (10.8) is finite if the matrix F^t_s has all its eigenvalues inside the unit circle. This is the case if for each $s \in T$, F_s has all its eigenvalues inside the unit circle and (F_s, H_s) is observable. Note, however, that these conditions are only sufficient, not necessary, for Γ^t_s to be finite, since Eq. (10.10) may be finite even if Ψ^t_s, obtained as the limit value of Eq. (10.8) is not finite.

Theorem 10.1. *For each $s \in t$, let the corresponding model be stable and observable and let $n_0 = -\infty$. Then the residuals sequences $(z_n - \hat{z}_{s,n})$; $s \in T$; $n \geq 0$ are ergodic.* Δ

Note that the stability and observability of the models, assumed in Theorem 10.1, are only sufficient, not necessary. In fact, we have proved ergodicity of the state residuals $(x_n - \hat{x}_{s,n})$, which is not a necessary condition to show the ergodicity of $(z_n - z_{s,n})$. In the sequel we shall directly use the following condition.

(c10.2) For each $s \in T$, the residuals sequence $(z_n - \hat{z}_{s,n})$ is ergodic.

10.3 Information, convergence and consistency

Consider the system Eq. (10.1) and the model set Eq. (10.2). Let condition (c10.1) hold. Then the conditional probability density of

z_n, given the past observations Z^{n-1}, corresponding to each model is

$$f_s(z_n|Z^{n-1}) = [(2\pi)^l|\Sigma_s|]^{-1/2}\exp\left\{-\frac{1}{2}(z_n - \hat{z}_{s,n})^T\Sigma_s^{-1}(z_n - \hat{z}_{s,n})\right\};$$

$$s \in T \tag{10.11}$$

where l is the dimension of z_n.

In Section 10.2, we defined for each pair $s, t \in T$

$$I_n(t;s) \equiv E_* \log \frac{f_t(z_n|Z^{n-1})}{f_s(z_n|Z^{n-1})}$$

and

$$d_n(t; s \equiv)|I_n(t;s)|$$

We have, for each $s \in T$

$$E_* \log f_s(z_n|Z^{n-1}) = -\frac{1}{2}\log 2\pi - \frac{1}{2}\log|\Sigma_s| - \frac{1}{2}\operatorname{tr}\Sigma_s^{-1}\Gamma_s^* \tag{10.12}$$

and for each pair $s, t \in T$

$$I_n(t;s) = I(t;s) = \frac{1}{2}\log|\Sigma_s| - \frac{1}{2}\operatorname{tr}\Sigma_s^{-1}\Gamma_s^* - \frac{1}{2}\log|\Sigma_t| - \frac{1}{2}\operatorname{tr}\Sigma_t^{-1}\Gamma_t^* \tag{10.13}$$

Let

$$L_s^r \equiv \log|\Sigma_s| + \operatorname{tr}\Sigma_s^{-1}\Gamma_s^r; \quad s, r \in T \tag{10.14}$$

Then we have

$$I_n(t;s) = \frac{1}{2}(L_s^* - L_t^*); \quad s, t \in \zeta \tag{10.15}$$

Also, by Eq. (2.27)

$$I_n(^*;s) \geq 0 \text{ for each } \quad s \in \zeta$$

Hence,

$$d(^*;s) = d(^*;s)_n = I_n(^*;s) \text{ for each } s \in \zeta$$

Thus, for any $s, t \in \zeta$

$$d(*; s) - d(*; t) = I_n(*; s) - I_n(*; t) = I_n(t; s) = \frac{1}{2}(L_s^* - L_t^*) \quad (10.16)$$

Hence,

$$d(*; s) \geq d(*; t)$$

if and only if

$$L_s^* \geq L_t^*$$

Theorem 10.2. *Let* (z_n) *be generated by Eq.* (10.1) *and let Condition* (c10.1) *hold.*

Then, under Condition (c10.2), *for any* $s, t \in \zeta$

$$\lim_{n \to \infty} h_t^s(Z^n) = 0 \ a.s. \quad (10.17)$$

if

$$L_t^* < L_s^* \quad (10.18)$$

and only if

$$L_t^* \leq L_s^* \quad (10.19)$$

Δ

Now let $\zeta = K \equiv (1, \ldots, p)$, i.e., let the parameter set be finite. Suppose that for some $t = k \in K$ we have Eq. (10.17) for each $s = j \in K$; $j \neq k$. Then it follows that ML estimates on the set M given by Eq. (10.2) will converge a.s. to the k-th model. Suppose that MAP and LS estimates are also defined for the set (i.e., assuming that the true system is a member of M) but that, in fact, the true system may not be a member of the model set. It has been shown (Ya and Tsypkin, 1968) that under Eq. (10.17), the MAP estimates will converge a.s. and the LS estimates will converge a.s. and in the mean-square to the k-th system.

Consider the following condition:

(c10.3) There exists some parameter $k \in K$ such that

$$L_k^* < L_j^* \quad \text{for all } j \in K;\ j \neq k \quad (10.20a)$$

The following result then follows immediately from Theorem 10.2.

Theorem 10.3. *Consider the system Eq.* (10.1) *and the model set* M, *given by Eq.* (10.2) *with* $\zeta = K$. *Let conditions* (c10.1) *and* (c10.2) *hold. Then under condition* (c10.3), *the ML, the MAP and the LS identification methods will converge a.s. and the LS identification method will also converge in the mean-square to the model set* M *specified by Eq.* (10.2). *Note that by Eq.* (10.16), *the identified model is the closest to the true model in the information metric* d. Δ,□

Corollary 10.1. *The convergence specified in Theorem* 10.3 *will be to a model in the model set* M *such that*

$$|L_k^* - L_*^*| = \min\{|L_j^* - L_*^*|; j \in K\} \tag{10.20b}$$

Δ

The identification methods will then converge to a parameter in K, closest to the true parameter in the scalar L, which, in turn, implies closeness of the corresponding models in terms of their output statistics. Note that

$$L_*^* = \log|\Sigma_*| + l \tag{10.21}$$

In the identification problem, the true parameter is unknown. If the true parameter can be assumed to belong to the parameter set, then Eq. (10.34) will have to be checked for all pairs of parameters in the set, i.e.,

(c10.4) For all pairs $i, j \in K, i \neq j$

$$L_j^i \neq L_i^i = \log|\Sigma_i| + l \tag{10.22}$$

The following result follows directly from condition (c10.3) and Corollary 10.1.

Theorem 10.4. *Let the system Eq.* (10.1) *belong to the set* M (*Eq.* (10.2)), *and let conditions* (c10.1) *and* (c10.2) *hold. Then, under Condition D, the true model is identifiable a.s. by the ML and MAP estimates and identifiable a.s. and in the m.s. by the LS estimate.*

10.4 L_1 convergence

We have shown in Section 9.4 the L_1 convergence of the likelihood ratios and the *a posteriori* probability ratios under condition (c9.1).

Furthermore, it was shown that under condition (c9.5), bounds on the L_1 convergence rates can be established, thus providing a measure of performance of the *ML* and the *MAP* estimation methods. We now show L_1 convergence and derive L_1 convergence bounds for the identification of stationary linear systems addressed in the present chapter.

Consider the system in Eq. (10.1) and the model set in Eq. (10.2) and let condition (c10.1) hold. We have shown (Eq. (10.17)) that under condition (c10.2)

$$\bar{I}_n(k;j) = \bar{I}(k;j) = \frac{1}{2}(L_j^* - L_k^*) \quad \text{for all } n$$

For each pair $k, j \in K$where $\mathrm{L}_j^*; j \in K$ are constants.

Theorem 10.5. *Consider the system in Eq.* (10.1) *and the model set M in Eq.* (10.2). *Under conditions* (c10.1) *and* (c10.3), *for each $j \in K;\ j \neq k$, the sequences $(h_j^k(Z^n))$ and $\left(\frac{\mathrm{f}^b(k|Z^n)}{\mathrm{f}^b(j|Z^n)}\right)$ converge in L_1 to infinity. Furthermore, the sequences converge at rates no slower than exponential.* Δ

The rates $\alpha_j = \bar{I}(k;j);\ j \in K; j \neq k$ can only be computed, as discussed in Section 5.2, if the true model is known. If the true model is only known to belong to the set M, then the rates can be bounded as follows. We have seen that if $k \in K$ is the true parameter, then (see Eq. (10.21))

$$L_k^* = \log|\Sigma_k| + \ell$$

Now, since

$$\alpha_j = \frac{1}{2}(L_j^* - L_k^*) \quad \text{for each } j \in K;\ j \neq k$$

where k is the true parameter, we have

$$\alpha_j = \frac{1}{2}(L_j^* - \log|\Sigma_k| - \ell)$$

Then

$$\alpha_j \geq \alpha \equiv \min\left\{\min\left\{\frac{1}{2}(L_j^k - \log|\Sigma_k| - \ell); L_j^k - \log|\Sigma_k| - \ell = 0\right\}\right\}$$

$$\text{for all } j \in K; j \neq k;\ k \in K \tag{10.23}$$

Equation (10.23) reads as follows: For each $k \in K$, suppose that k is the true parameter. If

$$L_j^k - \log|\Sigma_k| - \ell \geq 0 \text{ for all j } \in \text{K; j} \neq \text{k} \tag{10.24}$$

then take the min over j of Eq. (10.24). Continue the procedure over all $k \in K$, discarding such k for which Eq. (10.24) does not hold (since then k cannot be the true parameter, for which Eq. (10.24) always holds). Then take the least of all the minimum values of Eq. (10.24) found above. Note that this procedure does not identify the true parameter, but, rather, finds a lower bound for α_j over $j \in K$.

10.5 Model selection

In practice, when a mathematical model of a dynamical system is required for purposes of estimation and control, one often knows, to certain approximation, what the model should be. However, because of implementation constraints, one has to select a different model. Such is the case when the actual system is of high-order, but the available computation and storage capabilities are such that only a low-order model can be handled. Another modeling problem arises when the actual system's model is known to take its values, which may be time-varying, from a given set, but only a single model can be considered. An example of practical significance is the dynamical model of an aircraft, whose parameters vary considerably over its flight envelope. However, the airborne computation and storage capabilities are limited, and a single, or perhaps a finite, number of models of the aircraft dynamics must be used throughout its operation (Athans *et al.*, 1977). These are not identification problems in the strict sense. Nevertheless, the analysis in the previous sections suggests a natural extension of the results to the model selection problems introduced above. It should be emphasized that, unlike the investigation of convergence and consistency of parameter estimates, the following discussion applies to infinite and even noncompact parameter sets.

Suppose that the true system, or an approximate model of it, is known, but its dimension is too high for implementation of estimation and control procedures. A model of lower dimension is then desired. Let the true system, or an approximate model of it, be given by (10.1), and let

$$M \equiv (F_s, G_s, H_s, Q_s, R_s);\ s \in \zeta \tag{10.25}$$

be a model set of dimension lower than that of Eq. (10.1). The system coefficients in M depend on a parameter vector s belonging to a parameter set S. It is desired to find the model in the set M, which is closest to the true system $(F_s, G_s, H_s, Q_s, R_s)$ in some meaningful distance sense. The information distance approach (see Chapter 8) clearly suggests that the selected lower dimensional model is one that satisfies the criterion

$$\min_s \{L_s^*;\ s \in \zeta\} \tag{10.26}$$

Next, suppose that the model of a linear system whose parameters may be time-varying is known to take its values from a set represented by Eq. (10.25).

Two different cases may be considered:

1) The model takes a certain constant value in the set M, and there is no prior knowledge, even in a probabilistic sense, on what value it might be.
2) During the system's operation, its mathematical model varies over the model set M. However, the model's variation is unknown or is impossible to handle because of storage or computation limitations.

In either case it is desired to select a single model from the set M to represent the system throughout its operation. One criterion for the selection of such a model is that the maximum possible distance d between the representative model and the true model (whatever it might be) will be minimal.

The procedure for selecting the representative model from M will then be as follows. First, for each parameter $s \in S$, find the parameter t whose distance from s is maximal and the corresponding maximal distance. Then find the parameter s for which the maximal distance found in the first step is minimal.

The distance between a parameter s and the parameter t of the set S is maximized over t by maximizing with respect to t the following term

$$L_s^t = \log|\Sigma_s| + \operatorname{tr}\Sigma_s^{-1}\Gamma_s^t \tag{10.27}$$

where, as before

$$\Sigma_s \equiv E_s\{(z_n - \hat{z}_{s,n})(z_n - \hat{z}_{s,n})^T\}$$

is obtained by running a Kalman filter to steady-state, and

$$\Gamma_s^t \equiv E_t\{(z_n - \hat{z}_{s,n})(z_n - \hat{z}_{s,n})^T\}$$

is obtained by running a Lyapunov equation to steady-state.

The representative model is then found by solving the min-max problem

$$\min_s \max_t \{L_s^t; s, t \in \zeta\} \tag{10.28}$$

10.6 Concluding remarks

The procedures described in this chapter find, in general, a model in the model-set whose output statistics are best matched with those of the true system. However, for the modeling problem considered, the role of the output can be played by any linear function of the state variables. If, for instance, it is desired to emphasize certain variables that affect the system's performance more than the others, or that can be measured better than the others, then these variables can be selected as outputs for the model selection procedures.

The problem of selecting a single model from a model set can be generalized to a problem of selecting a number of models from the set, so that the model set is approximately represented by a finite set of models. An identification procedure can then be employed "on-line" to find the model in the finite set, which is closest to the true system. The selection of a finite model set would require, as a first step, the division of the infinite parameter set into a finite number of subsets. The way in which the parameter space should be divided would depend on considerations of the physical problem involved, but it seems obvious that the division would employ the metric topology of the parameter space (just as interval length is used in R^n to divide a rectangle into equal parts).

Chapter 11

Single and Multiple Stochastic Model Reduction

11.1 Introduction

Mathematical models, defined by a structure and sets of parameter variations or uncertainties, may be simplified by both structure reduction and parameter set reduction. First, the approximation of high-order and time-varying linear Gaussian models by low-order and time-invariant ones is considered. Here, following an earlier work (Baram and Be'eri, 1981), the proposed approach is based on maximizing the probabilistic ambiguity between the actual model and the approximate one and is applicable to general stochastic linear models. The set reduction criteria give rise to a min-max optimization problem and a min-max-min problem, which is converted to a constrained min-max problem. The algorithmic solution of the optimization problems is considered in detail, along with several approximation and discretization schemes. The application and the validity of the proposed approach are examined in comparison to traditional design considerations by solving numerical examples for several structure and set reduction problems.

Order reduction methods for essentially deterministic, time-invariant linear models have been studied extensively (see, e.g., (Genesio and Milanese, 1976)). Various techniques, ranging from dominant eigenvectors (Davison, 1966), canonical forms (Arumagan and Ramamoorty, 1973) and output deviation (Sinha and Bereznai, 1971), approximations of state-space models to continued fraction

(Wright, 1973), state feedback (Shaked and Karcanias, 1976), and Pade approximations of transfer functions (Shamash, 1975), have been suggested. Stochastic model reduction has been considered in rather general terms (Hawkes and Moore, 1976). On the other hand, with very few exceptions, e.g., (Nostrati and Meadows, 1973), where only response to initial conditions is considered, the problem of representing time-varying models by time-invariant ones does not seem to have attracted much interest. Neither has the problem of representing models with large parameter variation or uncertainty by a single model or a small model group received a systematic treatment. This is somewhat surprising, since simplified modeling of systems, operating under varying conditions is needed in many applications (e.g., in aerospace control and navigation systems) for analysis and design purposes. In this chapter, following an earlier work (Baram and Be'eri, 1981), we present a unified approach to the model simplification and reduction problems. The proposed approach is based on maximizing the expected log-likelihood function or the probabilistic ambiguity between the actual system and the simplified model. An approximation coefficient, which proves to be very useful in the analysis, is also defined. Employing state-space models in the time domain makes it possible to consider a general class of linear model approximation problems. In particular, a single design would take account of stochastic disturbances, deterministic inputs, and initial conditions, which have been treated previously in the order reduction literature by specialized techniques. For the case of time-invariant systems, transient response and steady-state approximations are naturally defined in the time domain and prove to be compatible with traditional dominant pole approximations. The problems of (order and time) structure reduction and of (parameter) set reduction give rise to different programming problems. Structure reduction is achieved by solving a min-max problem and set reduction to a model group requires solving a min-max-min problem. The algorithmic aspects of these problems are discussed in detail. The validity and the usefulness of the proposed techniques are demonstrated by low-order numerical examples. The resulting designs, based on our purely statistical criterion, are found to be in complete agreement with what would be obtained from system-theoretic considerations and "classical" designs in the frequency domain.

11.2 Structure (order and time) reduction

11.2.1 *The approximation criteria*

1) General: Consider a finite sequence of observations $Y^N = (y_1, \ldots, y_N)$ with $y_i \in R^l$. Let $\bar{M} = \{M(s); s \in S\}$ denote a set of arbitrarily simple (hence, not necessarily accurate) models for Y^N, depending on a parameter s. A widely accepted criterion for selecting a model for Y^N from $\bar{M}$ is the maximum likelihood criterion

$$\max_{s \in S} \log f_s\left(Y^N\right) \tag{11.1}$$

where $f_s(Y^N)$ is the probability density function of Y^N on $\Pi_{i=1}^N R^m$, corresponding to $M(s)$. Now suppose that the probability distribution of the observations is known. Then a reasonable criterion for selecting an approximate model for Y^N from $\bar{M}$ is that the expected value of the log-likelihood function is maximized, i.e.,

$$\max_{s \in S} \int f\left(Y^N\right) \log f_s\left(Y^N\right) dY^N = \max_{s \in S} E\left\{\log f_s\left(Y^N\right)\right\} \tag{11.2}$$

where E denotes expectation taken with respect to the known distribution of Y^N, $f(Y^N)$. As noted previously (Schweppe, 1973), the integral in Eq. (11.2) may be viewed as a version of the ambiguity function used in radar applications (Woodward, 1980). We call it here the probabilistic ambiguity between the actual system and the approximating model. It should be noted that in the above definition, it is assumed that when y_i has a well-defined probability density on R^l (i.e., that its probability measure is either discrete or absolutely continuous with respect to the Lebesgue measure). This is suitable for the Gaussian case treated in the sequel. When this is not the case (i.e., when Y may take values from mixed continuous and discrete sets), the ambiguity function is ill-defined. Instead, the approximation criterion would be defined by the Kullback–Leibler's information (Kullback and Leibler, 1951) between the corresponding probability measures, i.e.,

$$\max_{s \in S} \int \left[\log \frac{dP_s\left(Y^N\right)}{dP\left(Y^N\right)} \right] dP\left(Y^N\right)$$

assuming that the two measures are mutually absolutely continuous. It is not difficult to see that whenever these measures are absolutely continuous with respect to the Lebesgue measure, the two criteria coincide. We shall find it convenient, without losing generality, to write the criterion in the form

$$\min_{s \in S} J(s) \tag{11.3}$$

where

$$J(s) = -\frac{2}{N} E \log f_s\left(Y^N\right)$$

In order to evaluate the compatibility of the selected approximate model, it is reasonable to compare the value of the criterion function for the selected model with the one corresponding to the given distribution. Furthermore, a good measure should be confined to a specified range, so as to provide comparative information on different approximations. Following an enlightening comment (Tugnait, 1982), we define

$$J'(s) = -\frac{2}{N} E\left\{\log\left[f_s\left(Y^N\right) / f_v\left(Y^N\right)\right]\right\} \tag{11.4}$$

and

$$J'(*) = -\frac{2}{N} E\left\{\log\left[f\left(Y^N\right) / f_v\left(Y^N\right)\right]\right\} \tag{11.5}$$

yielding the approximation coefficient

$$K = \frac{J'(s)}{J'(*)} \tag{11.6}$$

Clearly, K measures the difference between the approximate model and the actual one.

2) Linear Gaussian systems:

Consider a linear model in state-space form

$$\begin{aligned} x_{n+1} &= F_n x_n + D_n u_n + G_n w_n \\ y_n &= H_n x_n + v_n \quad n = 1, \ldots, N \end{aligned} \tag{11.7}$$

where x_n, y_n, and u_n are k, l, and m-dimensional, respectively, x_0 is normally distributed with mean ξ and covariance ψ, (u_n) is a known sequence, and (w_n) and (v_n) are mutually uncorrelated, zero-mean white noise sequences with covariance (Q_n) and (R_n), respectively. We denote the given model by

$$M = \{(F_n, G_n, D_n, H_n, \xi_n, \psi_n, Q_n, R_n)\,;\, n = 1, \ldots, N\} \tag{11.8}$$

A time-invariant model of the form

$$\begin{aligned} x_{n+1} &= F x_n + D u_n + G w_n \\ y_n &= H x_n + u_n \quad n = 1, \ldots, N \end{aligned} \tag{11.9}$$

that approximates M in some sense is required. This simplified model may also be of order less than k. It is reasonable, although not essential to our analysis, to restrict this simplified model to some standard (e.g., standard controllable) form. The simplified model set is denoted by

$$\bar{M} = \{M_s = (F_s, G_s, D_s, H_s, \xi_s, \psi_s, Q_s,\, R_s)\,;\, s \in S \in R\} \tag{11.10}$$

where s is the vector of free parameters of the model and S is the parameter set. In order to write the approximation criterion for this problem, we first note that

$$\log f_s\left(Y^N\right) = \sum_{n=1}^{N} \log f_s\left(y_n \left| Y^{n-1}\right.\right) \tag{11.11}$$

where $f_s(Y^N)$ denotes the conditional probability density corresponding to M_s. For the normal models concerned, we have

$$\begin{aligned} f_s\left(y_n \left| Y^{n-1}\right.\right) = {} & \left[\left(2\pi^l\right)\left|\Sigma_{s,n}\right|\right]^{-1/2} \exp \\ & \times \left\{-1/2\left(y_n - \hat{y}_{s,n}\right)^T \left|\Sigma_{s,n}^{-1}\right| \left(y_n - \hat{y}_{s,n}\right)\right\} \end{aligned} \tag{11.12}$$

where

$$\hat{y}_{s,n} = E_s \left\{ y_n \left| Y^{n-1} \right. \right\} \tag{11.13}$$

and

$$\Sigma_{s,n} = E_s \left\{ \left(y_n - \hat{y}_{s,n} \right) \left(y_n - \hat{y}_{s,n} \right)^T \right\} \tag{11.14}$$

Denoting

$$J_n(s) = \log |\Sigma_{s,n}| + \mathrm{tr}\Sigma_{s,n}^{-1}\Gamma_{s,n} + \log (2\pi)^l \tag{11.15}$$

where

$$\Gamma_{s,n} = E \left\{ \left(y_n - \hat{y}_{s,n} \right) \left(y_n - \hat{y}_{s,n} \right)^T \right\} \tag{11.16}$$

we have

$$J(s) = \frac{1}{N} \sum_{n=1}^{N} J_n(s) \tag{11.17}$$

The first and the second derivatives of $J(s)$ with respect to s, required for a gradient-type minimization procedure, are provided in the Appendix of this chapter.

3) Time-invariant systems: In reducing the order of time-invariant systems, two types of approximations may be considered in the time domain:

a) transient response or finite time design and
b) steady-state design.

When the system's operation is given on a finite time length, or when emphasis of the transient response characteristics is desired in the design, the approximation criterion function is given by Eq. (11.17), where N is the given (sampled) time length, the duration of the transient response, or an arbitrary initial time period selected for design purposes. Characteristic response times may be determined by eigenvalue analysis or simulation of the original system.

When an emphasis of the steady-state characteristics of the original system is required in the reduced-order design, the criterion function takes its steady-state value, i.e.,

$$J(s) = \lim \frac{1}{N} \sum_{n=1}^{N} J_n(s) = \log |\Sigma_s| + \mathrm{tr}\Sigma_s^{-1}\Gamma_s + \log(2\pi)^l = J_{s.s.} \tag{11.18}$$

where Σ_s, Γ_s are the steady-state values of $\Sigma_{s,n}, \Gamma_{s,n}$, respectively.

11.2.2 *Special cases and examples*

We now turn to examine the usefulness of the proposed approach by solving specific low-order examples for a variety of model simplification problems. We first consider the problem of order reduction for time-invariant models, using steady-state and initial response design procedures. The effect of a deterministic input on the design is also examined. We then consider the approximation of a time-varying system by a time-invariant model, and, finally, the approximation of a time-varying system by a lower-order, time-invariant model is demonstrated.

Example 11.1. Given a second-order system

$$\begin{aligned} x_{n+1} &= \begin{bmatrix} 0 & 1 \\ -a & 1 \end{bmatrix} x_n + \begin{bmatrix} 0 \\ 1 \end{bmatrix} w_n \\ y_n &= \begin{bmatrix} 1 & 0 \end{bmatrix} x_n + v_n \quad R = Q = 1 \end{aligned} \tag{11.19}$$

It is desired to find a first-order model of the form

$$\begin{aligned} x_{n+1} &= s x_n + w_n \\ y_n &= x_n + v_n \quad R_s = Q_s = 1, \quad s \in S = (-\infty, \infty) \end{aligned} \tag{11.20}$$

that approximates the given model. Using our approximation criterion, it is necessary to find the value of s that maximizes the function $J(s)$. We seek a steady-state approximation and an initial response approximation for the first five sampling times. "Classical" design considerations imply that the frequency response of the approximate

Table 11.1. Approximation coefficient values for Example 11.1's steady-state design (a, unchanged, $s, K^{(3)}$ (Baram and Be'eri, 1981), $K^{(1)}$, $K^{(2)}$ (Tugnait, 1982)).

Case	a	s	$k^{(1)}$	$k^{(2)}$	$k^{(3)}$
1	0.0099	0.99	0.9999	0.9999	0.9999
2	0.09	0.91	0.9991	0.9994	0.9985
3	0.16	0.85	0.9974	0.9968	0.9961
4	0.25	0.78	0.9934	0.9878	0.9987
5	0.5	0.62	0.9651	0.9073	0.9340
6	0.9	0.50	0.6314	0.6400	0.4627

model should be similar to that of the original system in the design range. The optimal values of the parameter s and the corresponding approximation measure K for several values of a are given in Table 11.1. The poles of the given system and the approximating models in the z-plane are shown in Fig. 11.1.

It is seen that when the original system has a dominant (low-frequency) pole (Case 1), both designs give models whose poles are close to the dominant pole. In particular, the pole of the steady-state approximation is practically identical to the low-frequency pole, while the pole corresponding to the initial response design is of higher frequency, as expected. As the poles of the original system come closer (Cases 2–4), the pole of the approximate model lies to their right, for both designs, to give similar frequency responses to that of the original poles, in the respective ranges. (Note again, that the steady-state design pole lies to the right of the initial response design pole.) Note that for Cases 1–4, the approximation coefficient assumes higher values for the steady-state design, indicating better approximation. When the poles of the original system are complex (Cases 5 and 6), each of the approximate models has a real pole with a lower real frequency than the complex couple, giving similar frequency response characteristics. As the imaginary part of the original system's poles increases, the approximate model's poles converge to the real part of the original system's pole. Note that here the approximation coefficients for the steady-state design decreases

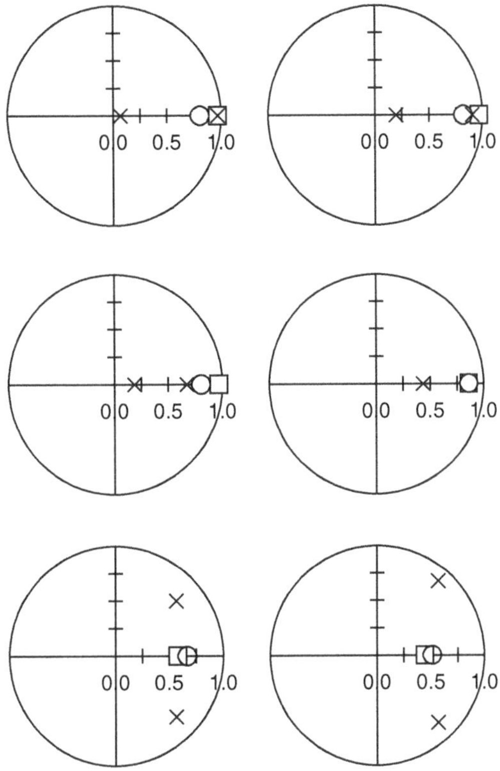

Figure 11.1. z-plane poles of second-order model and first-order approximation: original model pole $\times$, approximate model pole $\Box$ in steady-state design, and $\bigcirc$ in initial response design.

drastically (Case 6), while its value for the initial response design remains relatively high. This means that when the original system has oscillatory characteristics (complex poles), a transient response approximation is more suitable than a steady-state approximation.

Next consider the system given by Eq. (11.10) with $a = 0.25$, but let the reduced order model be

$$x_{n+1} = s_1 x_n + w_n$$

$$y_n = s_2 x_n + v_n$$

$$Q_s = s_3 \quad R_3 = s_4$$

A steady-state design is desired. The resulting parameter values were found to be: s1 = 0.6785, s2 = 1.203, s3 = 1.299, and s4 = 0.467 with the approximation coefficient K = 0.9993. As might have been expected, the approximation here is better than in the previous design (Case 1), as effectively all the model's parameters were free to take their optimal values of R_s.

Finally, consider the system given by (11.20), with an additional deterministic input, i.e.,

$$x_{n+1} = \begin{bmatrix} 0 & 1 \\ -a & 1 \end{bmatrix} x_n + \begin{bmatrix} 0 \\ 1 \end{bmatrix} u_n + \begin{bmatrix} 0 \\ 1 \end{bmatrix} w_n$$

$$y_n = \begin{bmatrix} 1 & 0 \end{bmatrix} x_n + v_n \quad R = Q = 1$$

The desired reduced-order model is of the form

$$x_{n+1} = s x_n + u_n + w_n$$

$$y_n = x_n + v_n \quad R_s = Q_s = 1, \;\; s \in S = (-\infty, \infty)$$

with

$$u_n = 1$$

For $a = 0.25$ and a steady-state design, we get $s = 0.75$ and $K = 0.9856$. Comparing these results to Case 4 above, we see that additional deterministic input has a rather marginal effect on the pole selection and approximation quality.

Example 11.2. Given a time-varying system

$$x_{n+1} = F_n x_n + w_n$$

$$y_n = x_n + v_n$$

$$\xi_0 = 0 \quad \psi_0 = 0$$

$$R_n = Q_n = 1$$

consider the following cases:

a) *"Fading Memory"*:

$$F_n = 1 - \frac{n}{10} \quad n = 1, \ldots, 10$$

b) *"Intensifying Memory"*:

$$F_n = \frac{n}{10} \quad n = 1, \ldots, 9$$

Time-invariant models of the form

$$x_{n+1} = sx_n + w_n$$

$$y_n = x_n + v_n$$

$$\xi_0 = 0 \quad \psi_0 = 0$$

$$R_n = Q_n = 1$$

that approximate the given time-varying systems are desired. It is reasonable to expect the resulting parameter values to be close to the average (0.45 in both cases) tending, on one hand, towards later F_n values (due to the fading past effect of a stable system) and, on the other, towards larger F_n values (which have a greater effect on the output covariance as they bring the system closer to instability). This means that the selected value of the parameter s should be greater for the intensifying memory case.

The resulting values of the parameter s and the approximation coefficient are given below.

Case	s	K
a	0.51	0.980
b	0.57	0.986

It is seen that in both cases, average values were selected. The influence of greater F_n values is seen from the fact that in both cases $s > 0.45$. This effect is stronger for the case of intensifying memory, as expected. Also, note that the approximation coefficient is somewhat higher (i.e., the approximation is somewhat better) in the latter case, which might be explained by the fact that the system's dominant characteristics are somewhat more definite in the latter case.

Example 11.3. Given the second-order time-varying system

$$x_{n+1} = \begin{bmatrix} 0 & 1 \\ 0.1n & 1 \end{bmatrix} x_n + \begin{bmatrix} 0 \\ 1 \end{bmatrix} w_n$$

$$y_n = \begin{bmatrix} 1 & 0 \end{bmatrix} x_n + v_n \quad n = 0, \ldots, 9$$
$$R = Q = 1$$

It is desired to find a first-order invariant model as in Example 11.2. The original system's eigenvalues are listed below.

n:	0	1	2	3	4	5	6	7	8	9
Eigenvalues:	1.00	1.09	1.17	1.24	1.31	1.37	1.42	1.47	1.52	1.57
	0.00	−0.09	−0.17	−0.24	−0.31	−0.37	−0.42	−0.47	−0.52	−0.57

Employing our representation criterion, we get for the approximate model $s = 1.45$ with $K = 0.9790$. Note that the selected s value tends towards later values of the "unstable" eigenvalue of the original system, as might have been expected. The fact that both the original system and the approximate model are unstable is of no major consequence here, since the system's operation and the approximation are required for the first 10 samples only.

11.3 Set reduction

11.3.1 *General*

It is sometimes desired to reduce a model set defined by a structure and a set of parameter values to a single model with fixed parameter values and the same or simpler structure that represents the model set in some sense. In certain cases, however, important characteristics (e.g., those dynamic characteristics that are particular to subsonic and supersonic flights in modeling aircraft dynamics) may be lost in a single model design. A finite group of representative models that "spans" the model set in some sense would then be desired.

In the previous section, we considered the problems of order and time structure reduction. We had a known stochastic model for the system and a model set from which a simpler approximate model was to be selected. This was achieved by minimizing a probabilistic ambiguity function with respect to a single variable parameter. For the model set reduction problem considered in the present section, the original model becomes a model set and the ambiguity function becomes a function of two variable parameters. The two parameter sets may or may not coincide, depending on whether structure reduction is also sought and on the set reduction procedure used.

11.3.2 *Set reduction criteria*

Let $S \in R^p$ and $T \in R^q$ be two parameter sets so that for each $s \in S$ and $t \in T$, there exist probability density functions $f_s(Y^N)$ and $f_t(Y^N)$ for an observation record $Y^N = (y_1, \ldots, y_N)$ on its space of definition. A scalar function is defined on the product set $S \times T$ by

$$J_N(t,s) = -\frac{2}{N} E_t \log f_s\left(Y^N\right) = -\frac{2}{N} \sum_{n=1}^{N} E_t \log f_s\left(Y^{N-1}\right)$$

$$= -\frac{2}{N} \sum_{n=1} J_n(t,s) \tag{11.21}$$

where E_t denotes expectation with respect to the density $f_t(Y^N)$. Note that $J_N(t,s)$ is the negative probabilistic ambiguity between the models corresponding to s and t. When the observations are ergodic (e.g., stationary Gaussian), $J(t,s)$ has a limit value

$$J(t,s) = \lim_{N \to \infty} J_N(t,s) = -2E_t \log f_s\left(Y_n \left| Y^{N-1}\right.\right) \tag{11.22}$$

Example 11.4. Consider a second-order linear Gaussian model set in state-space form

$$x_{n+1} = \begin{bmatrix} 0 & 1 \\ t & 1 \end{bmatrix} x_n + \begin{bmatrix} 0 \\ 1 \end{bmatrix} w_n$$

$$y_n = \begin{bmatrix} 1 & 0 \end{bmatrix} x_n + v_n$$

$$E\{w_n\} = E\{v_n\} = 0$$

$$\text{cov}\{w_n\} = \text{cov}\{v_n\} = 1$$

where

$$-0.25 \le t \le 0$$

and a first-order model set

$$x_{n+1} = s x_n + w_n$$

$$y_n = x_n + v_n$$

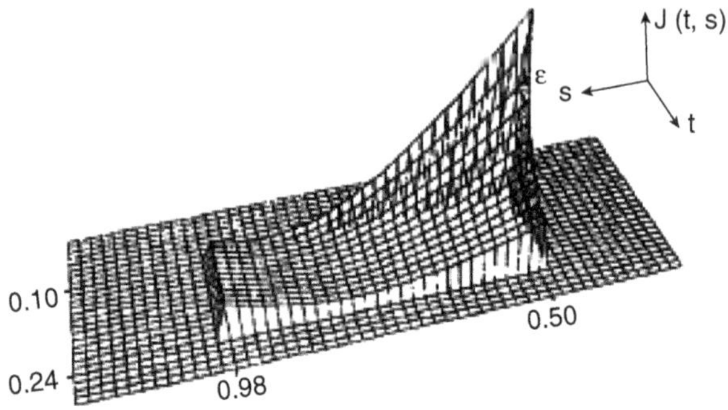

Figure 11.2. The function $J(t,s)$ for Example 11.4.

$$E\{w_n\} = E\{v_n\} = 0$$

$$\text{cov}\{w_n\} = \text{cov}\{v_n\} = 1$$

where

$$0.5 \leq s \leq 1.0$$

The function $J(t,s)$ for this example has been plotted using a three-dimensional plotting program and is given in Fig. 11.2.

If Y^N does not possess well-defined densities, $J(t,s)$ will be replaced by Kullback–Leibler's information between the corresponding probability measures

$$I(t,s) = \int_{Y^N} \log\left[\frac{dP_s(Y^N)}{dP_t(Y^N)}\right] dP_t(Y^N)$$

provided that P_s and P_t are mutually absolutely continuous for all $s \in S,\ t \in S$. We shall see that in contrast to the structure reduction problems treated in Section 11.2, the measures $J(t,s)$ and $I(t,s)$ do not coincide for the set reduction problem treated in the present section, even when $f_s(Y)$ and $f_t(Y)$ are well defined, and that there is an obvious computational advantage to using $J(t,s)$ when possible.

Given two parameter sets T and S and two model sets defined on them, the problem of concern here is to find a finite group of models defined on S that best represents the entire model set defined on T. We call T the set of definition (of the original model) and S the set of representation. S and T may, but need not, coincide and each may be continuous or discrete. When the original model is completely specified by a parameter value t, a reasonable criterion for selecting an approximate model from a model set defined on S is that the ambiguity between the actual model and the approximate one is maximized (see Section 11.2), i.e.,

$$\min_{s \in S} J(t, s) \tag{11.23}$$

When the actual model is not specified, but instead is given on a parameter set T, a reasonable criterion for selecting a single representative model defined on S is that the minimum possible ambiguity between the actual model (whatever it might be) and the selected one is maximized. The criterion is then

$$\min_{s \in S} \max_{t \in T} J(t, s) \tag{11.24}$$

Next, suppose that the model set defined on T is to be represented by a finite group of L models, defined on S. Denote the finite group of parameter values to be selected by

$$P = \left(s_j \in S' j = 1, \ldots, L\right)$$

and suppose that $t \in T$ is the actual parameter value. Then, using the criterion (11.23), the best representative parameter from P is the one giving

$$\min_{s_j \in P} J\left(t, s_j\right) \tag{11.25}$$

Now, P must be selected from S so that the model obtained by Eq. (11.25) is the best representation for T. The min-max criterion (11.24) requires that the selection satisfies

$$\min_{P \in S} \max_{t \in T} \left\{ \min_{s_j \in P} J(t, s) \right\} \tag{11.26}$$

We next derive a measure of compatibility for the selected model or model group by extending the measure K defined in Section 11.2 as follows. For set reduction to a single model, we obtain a min-max point s^0. Defining

$$t^0 = \arg\max_{t \in T} J\left(t, s^0\right)$$

and

$$K^{'}\left(s^0\right) = \frac{J\left(t^0, s^0\right) - J\left(t^0, t^0\right)}{J\left(t^0, s^0\right)}$$

it can be shown (as in Section 11.2) that

$$0 \le K^{'}\left(s^0\right) \le 1$$

$K^{'}(s^0)$ measures the (normalized) difference between the selected parameter s^0 and t^0, which is "the most different" from s^0. $K^{'}(s^0)$ then measures the difference between the selected model and the actual one in the "worst case". A measure of approximation is then defined as the complement

$$K\left(s^0\right) = 1 - K^{'}\left(s^0\right) = \frac{J\left(t^0, t^0\right)}{J\left(t^0, s^0\right)} \tag{11.27}$$

For the problem of set reduction to a finite group P^0, we similarly define

$$t^0 = \arg\max_{t \in T} \left\{ \min_{S \in P^0} J(t, s) \right\}$$

and

$$J\left(t^0, P^0\right) = \max_{t \in T} \left\{ \min_{s \in P^0} J(t, s) \right\}$$

Then a measure of difference between the selected model and the actual one for the "worst case" is

$$K^{'}\left(P^0\right) = \frac{J\left(t^0, P^0\right) - J\left(t^0, t^0\right)}{J\left(t^0, P^0\right)}$$

and a measure of approximation is defined by the complement

$$K\left(P^{0}\right)=1-K^{'}\left(P^{0}\right)=\frac{J\left(t^{0},t^{0}\right)}{J\left(t^{0},P^{0}\right)} \tag{11.28}$$

11.3.3 *The optimization problem*

The problem at hand is to construct procedures for solving the optimization problems (11.24) and (11.26). Min-max problems have received considerable attention in the literature, and several procedures can be adapted to Eq. (11.24). On the other hand, there does not seem to be anything in the literature concerning problems such as Eq. (11.26), which is a generalization of the min-max problem. The following procedure reduces the solution of Eq. (11.26) into iterative solutions of min-max problems under constraints.

Step 0: On the k-th iteration, have the group

$$P_k=(s_{k,j};\ j=1,\ldots,L).$$

Step 1: For each $s_{k,j}\in P_k$, find a subset $T(s_{k,j})\in T$, such that for $t\in T(s_{j,k})$

$$J\left(t,s_{k,j}\right)\geq J\left(t,s_{k,i}\right)\quad\text{for any}\ \ i\neq j.$$

Step 2: Find the min-max point in S for each $T(s_{k,j}), j=1,\ldots,L$, and obtain a new group $P_{k+1}=(s_{k+1,j},\ j=1,\ldots,L)$.

Step 3: Check a stopping condition. Set $k=k+1$ and repeat Steps 1 and 2 until the stopping condition is satisfied.

The above procedure is described in Fig. 11.3. The actual algorithm performs Steps 1 and 2 together as follows.

Step 1, 2: For each $s,j\in P_k$, $j=1,\ldots,L$

find

$$\min_{s\in S}\max_{t\in T}J(t,s) \tag{11.29}$$

Subject to

$$J\left(t,s_{k,j}\right)\geq J\left(t,s_{i,j}\right)\ \text{for any}\,i\neq j \tag{11.30}$$

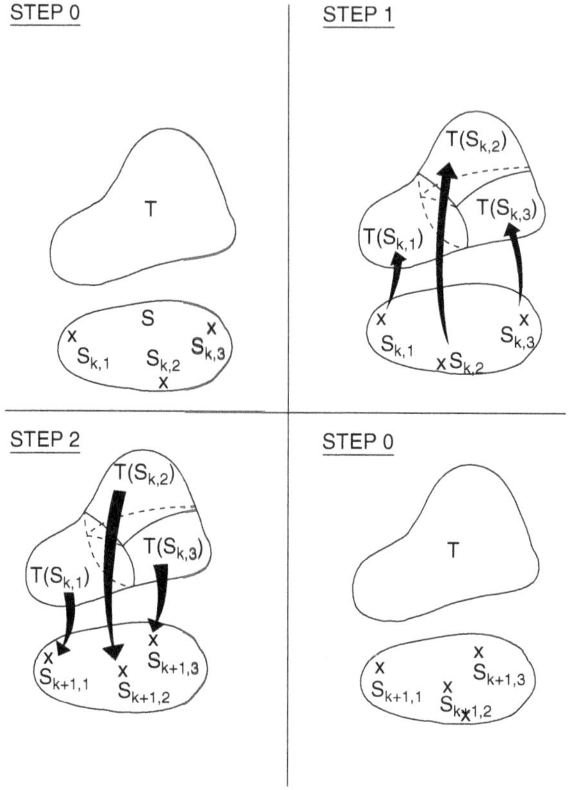

Figure 11.3. Iterative reduction of a continuous set to a finite group.

Note that the constrains involve only the parameter t and not s. It follows that in solving Eq. (11.29) in each iteration, only the minimization is performed under the constraint Eq. (11.30).

The solution of Eq. (11.24) is then a (considerably simpler) subproblem of the solution of Eq. (11.26). This should come as no surprise, as representation by a single model is a special case of representation by a finite group. The representation problems at hand involve solving unconstrained (in the case of a single representative model) or constrained (in the case of a representative group) min-max problems. These, in turn, involve unconstrained minimization and constrained or unconstrained maximization steps. Note that for the case of linear Gaussian models on continuous parameter sets, exact expressions for the first and the second derivatives of the criterion function $J(s,t)$ with respect to the parameter s are provided in the Appendix of this chapter and can be used in performing the minimization step on S. The derivatives with respect to the parameter t for the maximization step on T can be derived in a similar manner (replacing the original model by the model corresponding to t and differentiating with respect to t).

11.3.4 *Special cases*

1) General: Before addressing the optimization problems Eq. (11.24) and Eq. (11.26) for continuous parameter sets S and T, we consider the case where both sets are discrete and the case where one set is continuous and the other discrete. These cases may be viewed as valid optimization problems in their own right, applying to practical problems of selecting between finite alternatives or as approximations of the continuous parameter problem. Such simplified versions of the optimization problem are also used as intermediate iterative steps in certain algorithms for solving min-max problems on continuous parameter sets, e.g., (Salmon, 1968). We shall use (a, b) to denote the finite set composed of a and b and [a, b] to denote the closed interval between a and b.

2) Finite parameter sets: The parameter set T may be given as a finite set or constructed as such from a continuous set by some discretization, so as to approximate the original set. The set of representation S may coincide with T or define a finite set of simpler models (say, of lower order) than those defined by T. The solution of the min-max problem Eq. (11.24) is then found by simple comparison of $J(s,t)$ values of pairs s and t. The solution of Eq. (11.26) is found by the procedure proposed in the previous section, solving the min-max problem on each iteration.

Example 11.5. Given the model set

$$x_{n+1} = \mathrm{t}\mathrm{x}_n + \mathrm{w}_n \quad \mathrm{t} \in \mathrm{T}$$

$$y_n = x_n + \upsilon_n$$

$$x_0 = 0, \psi_0 = 1$$

$$R = Q = l$$

it is desired to find a single representative model from the set for the following cases:

a) steady-state design for $T = (0.1. 0.2. 0.3. 0.4. 0.5. 0.6)$,
b) steady-state design for $T = (0.1. 0.2,\ 0.3,\ 0.8. 0.9. 0.99)$,
c) initial response design (first 3 samples) for $T = (0.1.\ 0.2,\ 0.3,\ 0.8,\ 0.9,\ 0.99)$.

The representative models are found by applying the criterion Eq. (11.24) with $S = T$. The selected values of t^0 and the approximation measure $K\left(t^0\right)$ (which are given for comparison with later results) are listed below.

Case	t^0	$K(t^0)$
a	0.4	0.9836
b	0.99	0.8223
c	0.8	0.8945

It is seen that in Case a, an "average" value for the parameter was selected from the set. In Case b, however, the extremal value was selected. This is explained by the fact that the system associated with 0.99 is nearly unstable, and the resulting higher output covariance has greater influence on the criterion function $J(t, s)$ than the other models. This is a desirable result, as the nearly unstable models should be approximated more closely than the strongly stable ones. For the initial response design, Case c, the selected parameter value has shifted to the left, as the effect of near instability is small for the short time period.

Example 11.6. Now it is desired to find the best two representative models from the model set of Example 11.4 with $T = (0.1, 0.2, 0.3, 0.4, 0.5, 0.6)$ for a steady-state design. Starting with the initial parameters $P_0 = (0.2. 0.4)$, the constrained min-max problem is, on the first iteration,

$$\min_{s \in T} \max_{t \in T} J(s, t)$$

subject to (maximization step only)

$$J(t, 0.2) \leq J(t, 0.4)$$

and

$$\min_{s \in T} \max_{t \in T} J(s, t)$$

subject to

$$J(t, 0.4) \leq J(t, 0.2)$$

After computing $J(t, 0.2)$ and $J(t, 0.4)$ values for all $t \in T$, the above constrained min-max problems become

$$\min_{s \in T_1} \max_{t \in T_1} J(s, t) \quad T_1 = (0.1,\ 0.2,\ 0.3)$$

and

$$\min_{s \in T_2} \max_{t \in T_2} J(s, t) \quad T_2 = (0.4,\ 0.5,\ 0.6)$$

yielding

$$P_1 = (0.2,\ 0.5)$$

which remains unchanged in further iterations

3) Finite-continuous parameter sets: Intermediate approximations for continuous set reduction may be achieved by discretizing one of the parameter sets, yielding one of the following two situations:

a) discrete definition set and continuous representation set.
b) continuous definition set and discrete representation set.

Either discretization would reduce the computation effort involved in solving the continuous set reduction problem considerably, as the maximization step (Case a) or the minimization step

(Case b) of the min-max procedure are performed on finite sets. There does not seem to be an obvious preference to *a* or *b a priori* unless they are suggested by the given problem (for instance, Case a would arise in finite set reduction when structure reduction is also performed).

Optimization procedures for solving min-max problems when one parameter is given on a continuous set and the other on a discrete finite set have been suggested in the literature (Gupta and Mehra, 1974), and some are available in widely used subroutine libraries (Hooper, 1978). These subroutines may be used to solve our min-max problem (Eq. (11.34)). Note, however, that these general subroutines are tailored for problems where only the criterion function (and none of its derivatives) is available or where only the function and its first derivative are available. In our case, we have first and second derivatives available for linear Gaussian models, so that better solutions (faster convergence) may be achieved.

Example 11.7. It is desired to find a single representative from the set

$$x_{n+1} = \mathrm{s}\mathrm{x}_n + \mathrm{w}_n$$

$$y_n = x_n + v_n$$

$$x_0 = 0, \ \psi_0 = 1$$

$$R = Q = l$$

where $s \in S = [0.1, 0.6]$, the closed interval between 0.1 and 0.6. The set S is first "approximated" by a finite set, $T = (0.1, 0.2, 0.3, 0.4, 0.5, 0.6)$. Performing the optimization

$$\min_{s \in S} \max_{t \in T} J(t, s)$$

for a steady-state design, we obtain for the representative model $s^0 = 0.406$ and $K(s^0) = 0.9840$. Comparing these values to the ones obtained in Example 11.5 (Case a), we see that the selected values are very close, while the representation here is slightly better (larger K^0) than in the finite-finite case, as might have been expected.

Example 11.8. Let a discrete model set be given by

$$x_{n+1} = \begin{bmatrix} 0 & 1 \\ t & 1 \end{bmatrix} x_n + \begin{bmatrix} 0 \\ 1 \end{bmatrix} w_n$$

$$y_n = \begin{bmatrix} 1 & 0 \end{bmatrix} x_n + v_n \quad R = Q = 1$$

where $t \in T = (-0.09, \ -0.16, -0.25)$.

The eigenvalues of the system matrix are listed below.

t	Eigenvalues
−0.09	0.9; 0.1
−0.16	0.8; 0.2
−2.5	0.5; 0.5

It is desired to find a first-order representative model for the above set, i.e.,

$$x_{n+1} = sx_n + w_n$$

$$y_n = x_n + v_n \quad R = Q = 1$$

where s may take any value on the real line. Applying the min-max criterion, we get $s^0 = 0.842$. Note that the selected value is close to the higher eigenvalues of the original models, which is explained by the higher covariance values associated with the corresponding modes. From a frequency response viewpoint, the approximate representative model tends towards the low-frequency range of the original system due to the steady-state design.

4) Continuous parameter sets: A simple-minded procedure for solving min-max problems on continuous parameter sets is alternating minimization and maximization steps. Convergence can be normally guaranteed when the function has a saddle point (Dem'yanov and Malozemov, 1974). It can be argued, however, that, for the function $J(s,t)$, a saddle point will not normally exist. (Note that even in the very simple case of scalar parameters depicted in Fig. 11.2, a saddle point does not exist.) A procedure for solving min-max problems on continuous parameter sets by iterations on continuous discrete sets has been proposed (Salmon, 1968), showing convergence of the algorithm, even in the absence of a saddle point. This optimization method has been used in the following example.

Example 11.9. Let a first-order linear system be given by

$$x_{n+1} = tx_n + w_n$$

$$y_n = x_n + v_n \quad R_t = Q_t = l$$

where t takes its values from a given interval T on the real line. It is desired to find a single model that best represents the model set in steady-state. Applying our representation criterion Eq. (11.24) for two cases of T, the following values for t^0 were obtained.

Case	T	t^0
a	$[0.2, 0.5]$	0.370
b	$[0.1, 0.6]$	0.406

Note that the selected value of t^0 is greater than the mean value of the interval, which is explained by the fact that models with higher t values produce greater values of the output covariance (are "closer" to instability) and have greater influence on the criterion function. This effect is emphasized by the fact that although in Case b we have a symmetrical extension of the interval of Case a, the selected value of t^0 has shifted to the right.

Example 11.10. We now wish to find the best two and best three representative models for the continuous parameter model set of Example 11.9 with $T = [0.1, 0.6]$. The representation criterion is now the approximation criterion, Eq. (11.6). The selected parameter values are listed below along with the result obtained for a single representative model and the approximation measure, for comparison

One representative: (0.406); $K(s^0) = 0.984$

Two representatives: $(0.25, 0.50)$; $K(P^0) = 0.996$

Three representatives: $(0.2, 0.39, 0.54)$; $K(P^0) = 0.998$.

Note again, that the representations tend to emphasize the higher range of the parameter set for reasons discussed before. Also note that, as should be expected, the min-max approximation improves (higher K values) when more representative models are taken.

11.4 Conclusion

The problem of stochastic model simplification has been treated in the contexts of structure reduction and parameter set reduction. The approximation of high-order and time-varying systems by low-order and time-invariant models is achieved by maximizing the probabilistic ambiguity between the actual system and the selected model. The reduction of a model set defined on a set of parameter variation or uncertainty into a single model or a finite model group is achieved by finding the min-max and the min-max-min points of the ambiguity function. Set reduction may include structure reduction as a subproblem. The algorithmic aspects of the model simplification problem were considered in some detail. It should be emphasized that any efficient, general minimization, min-max and constrained min-max algorithms may be used to solve the respective model simplification problems. Several such algorithms have been tested and have shown good convergence properties for the examples considered. For linear Gaussian models, exact expressions for the gradient and the Hessian are computable so that better convergence rates can be achieved at the cost of more extensive computation per iteration. An efficient optimization procedure may employ a general algorithm, approximating the first or the second derivative, as a first stage, and a specialized algorithm, using the exact derivatives, as a second stage, for higher sensitivity in the neighborhood of the solution.

Several discretization and approximation schemes have been considered. The validity of the proposed theory and the applicability of the optimization techniques have been demonstrated by a variety of numerical examples, concerning linear-Gaussian models. The resulting designs have been shown to agree with the intuitive choice of "average" or "typical" models and traditional system-theoretic considerations.

Appendix A.11. Some supplements

The basic iteration in a gradient-type nonlinear programming method is

$$s^{k+1} = s^k - \rho_k H\left(s^k\right) \left.\frac{\partial J(s)}{\partial s}\right|_{s=s^k} \tag{A.11.1}$$

where s^k is the parameter vector at the k-th iteration, $H(s^k)$ is the inverse of the Hessian – the second partial matrix $(\partial^2 J(s)/\partial s^k|_{s=s^k})$ and ρ_k is the scalar step size parameter.

In order to apply the method to our problem, the gradient and Hessian of $J(s)$ must be derived. Clearly, these are obtained by summation of the gradient and Hessian of $J_n(S), n = 1, \ldots N$, which are derived in the sequel.

A. Gradient computation

From Eq. (11.15)

$$J_n(s) = \log |\Sigma_{s,n}| + \mathrm{tr}\Sigma_{s,n}^{-1}\Gamma_{t,n} + \log (2\pi)^l \tag{A.11.2}$$

using rules of matrix differentiation, we get

$$\frac{\partial J_n(s)}{\partial s_i} = \mathrm{tr}\left[\Sigma_{s,n}^{-1}\frac{\partial \Sigma_{s,n}}{\partial s_i}\right] + \mathrm{tr}\left[\Sigma_{s,n}^{-1}\frac{\partial \Gamma_{s,n}}{\partial s_i}\right] - \mathrm{tr}\left[\Sigma_{s,n}^{-1}\frac{\partial \Sigma_{s,n}}{\partial s_i}\Sigma_{s,n}^{-1}\Gamma_{s,n}\right] \tag{A.11.3}$$

where s_i is the i-th term of the parameter vector s. $\Sigma_{s,n}$ is obtained from

$$\Sigma_{s,n} = H_s \psi_{s,n} H_s^T + R_s \tag{A.11.4}$$

where $\psi_{s,n}$ is computed from the Riccati equation

$$\psi_{s,n+1} = F_s \psi_{s,n} F_s^T + C_s Q_s C_s^T - F_s K_{s,n} \Sigma_{s,n} K_{s,n}^T F_s^T \tag{A.11.5}$$

initialized at $\psi_{s,0} = \psi_s$ where $K_{s,n} = \psi_{s,n} H_s^T \left[H_s \psi_{s,n} H_s^T + R\right]^{-1}$ is the Kalman gain matrix.

The gradient of $\Sigma_{s,n}$ is obtained from Eq. (A.11.4) as

$$\frac{\partial \Sigma_{s,n}}{\partial s_i} = H_s \frac{\partial \psi_{s,n}}{\partial s_i} H_s^T + \frac{\partial R_s}{\partial s_i} + \frac{\partial H_s}{\partial s_i} \psi_{s,n} H_s^T + H_s \psi_{s,n} \frac{\partial H_s^T}{\partial s_i} \tag{A.11.6}$$

and using Eq. (A.11.5), we have

$$\frac{\partial \psi_{s,n+1}}{\partial s_i} = \tilde{F}_{s,n} \frac{\partial \psi_{s,n}}{\partial s_i} \tilde{F}_{s,n}^T + \Omega_{s_i} + \Omega_{s_i}^T \tag{A.11.7}$$

where

$$\tilde{F}_{s,n} \equiv F_s \left[I - K_{t,s} H_s\right]$$

and

$$\Omega_{s_i} \equiv \frac{\partial F_s}{\partial s_i} \psi_{s,n} \tilde{F}_s^T - F_s K_{s,n} \frac{\partial H_s}{\partial s_i} \psi_{s,n} \tilde{F}_s^T + \frac{1}{2} \frac{\partial}{\partial s_i} \left[G_s Q_s G_s^T\right]$$
$$+ \frac{1}{2} F_s K_{s,n} \frac{\partial R_s}{\partial s_i} K_s^T F_s^T$$

$\Gamma_{s,n}$ is obtained from

$$\Gamma_{s,n} = \bar{H}_s \left[\bar{\psi}_{s,n} + \bar{x}_{s,n} \bar{x}_{s,n}^T\right] \bar{H}_s^T \tag{A.11.8}$$

where $\bar{\psi}_{s,n}$ is obtained recursively from the equation

$$\bar{\psi}_{s,n+1} = \bar{F}_{s,n} \bar{\psi}_{s,n} \bar{F}_{s,n}^T + \bar{G}_{s,n} \bar{Q} \bar{G}_{s,n}^T \tag{A.11.9}$$

initialized at

$$\bar{\psi}_{s,0} = \begin{bmatrix} \psi_s & 0 \\ 0 & 0 \end{bmatrix}$$

and $\bar{x}_{s,n}$ is obtained from

$$\bar{x}_{s,n+1} = \bar{F}_{s,n} x_{s,n} + \bar{D}_s u_n \tag{A.11.10}$$

initialized at

$$\bar{x}_{s,0} = 0$$

We have used the notation

$$\bar{F}_{s,n} = \begin{bmatrix} F & 0 \\ F_s K_{s,n} H & F_s\left(I - K_{s,n} H_s\right) \end{bmatrix}; \quad \bar{G}_{s,n} = \begin{bmatrix} G & 0 \\ 0 & F_s K_{s,n} \end{bmatrix}$$

$$\bar{Q} = \begin{bmatrix} Q & 0 \\ 0 & R \end{bmatrix}; \; \bar{H} = \begin{bmatrix} H & -H_s \end{bmatrix}; \; \bar{x}_{s,n} \equiv E\begin{bmatrix} x_n \\ \hat{x}_{s,n} \end{bmatrix}; \; \bar{D}_{s,n} = \begin{bmatrix} D \\ D_s \end{bmatrix}$$

Denoting

$$\tilde{\psi}_{s,n} = \bar{\psi}_{s,n} + \bar{x}_{s,n}\bar{x}_{s,n}^T$$

the gradient of $\Gamma_{s,n}$ is

$$\frac{\partial \Gamma_{s,n}}{\partial s_i} = \bar{H}_s \frac{\partial \tilde{\psi}_{s,n}}{\partial s_i} \bar{H}_s^T + \frac{\partial \bar{H}_s}{\partial s_i} \tilde{\psi}_{s,n} \bar{H}_s^T + \bar{H}_s \tilde{\psi}_{s,n} \frac{\partial \bar{H}_s^T}{\partial s_i} \tag{A.11.11}$$

where

$$\frac{\partial \tilde{\psi}_{s,n}}{\partial s_i} = \frac{\partial \bar{\psi}_{s,n}}{\partial s_i} + \frac{\partial \bar{x}_{s,n}}{\partial s_i}\bar{x}_{s,n}^T + \bar{x}_{s,n}\frac{\partial \bar{x}_{s,n}^T}{\partial s_i}$$

where from Eq. (A.11.9)

$$\begin{aligned} \frac{\partial \bar{\psi}_{s,n+1}}{\partial s_i} &= \bar{F}_{s,n} \frac{\partial \bar{\psi}_{s,n}}{\partial s_i} \bar{F}_{s,n}^T + \frac{\partial}{\partial s_i}\left[\bar{G}_{s,n} Q \bar{G}_{s,n}^T\right] \\ &\quad + \frac{\partial \bar{F}_{s,n}}{\partial s_i} \bar{\psi}_{s,n} \bar{F}_{s,n}^T + \bar{F}_{s,n} \bar{\psi}_{s,n} \frac{\partial \bar{F}_{s,n}^T}{\partial s_i} \end{aligned} \tag{A.11.12}$$

and from Eq. (A.11.3)

$$\frac{\partial \bar{x}_{s,n+1}}{\partial s_i} = \bar{F}_{s,n} \frac{\partial \bar{x}_{s,n}}{\partial s_i} + \frac{\partial \bar{F}_{s,n}}{\partial s_i} \bar{x}_{s,n} + \frac{\partial \bar{D}_s}{\partial s_i} u_n \tag{A.11.13}$$

B. Hessian computation

From Eq. (A.11.3), we get

$$\frac{\partial^2 J_n(s)}{\partial s_j \partial s_i} = -\mathrm{tr}\left[\Sigma_{s,n}^{-1} \frac{\partial \Sigma_{s,n}}{\partial s_j} \Sigma_{s,n}^{-1} \frac{\partial \Sigma_{s,n}}{\partial s_i}\right] + \mathrm{tr}\left[\Sigma_{s,n}^{-1} \frac{\partial^2 \Sigma_{s,n}}{\partial s_j \partial s_i}\right]$$

$$
\begin{aligned}
&-\mathrm{tr}\left[\Sigma_{s,n}^{-1}\frac{\partial \Sigma_{s,n}}{\partial s_j}\Sigma_{s,n}^{-1}\frac{\partial \Gamma_{s,n}}{\partial s_i}\right]+\left[\Sigma_{s,n}^{-1}\frac{\partial^2 \Gamma_{s,n}}{\partial s_j \partial s_i}\right]\\
&+\mathrm{tr}\left[\Sigma_{s,n}^{-1}\frac{\partial \Sigma_{s,n}}{\partial s_j}\Sigma_{s,n}^{-1}\frac{\partial \Sigma_{s,n}}{\partial s_i}\Sigma_{s,n}^{-1}\Gamma_{s,n}\right]\\
&-\mathrm{tr}\left[\Sigma_{s,n}^{-1}\frac{\partial^2 \Sigma_{s,n}}{\partial s_j \partial s_i}\Sigma_{s,n}^{-1}\Gamma_{s,n}\right]\\
&+\mathrm{tr}\left[\Sigma_{s,n}^{-1}\frac{\partial \Sigma_{s,n}}{\partial s_i}\Sigma_{s,n}^{-1}\frac{\partial \Sigma_{s,n}}{\partial s_j}\Sigma_{s,n}^{-1}\Gamma_{s,n}\right]\\
&-\mathrm{tr}\left[\Sigma_{s,n}^{-1}\frac{\partial \Sigma_{s,n}}{\partial s_i}\Sigma_{s,n}^{-1}\frac{\partial \Gamma_{s,n}}{\partial s_j}\right] \qquad \text{(A.11.14)}
\end{aligned}
$$

For the calculation of Eq. (A.11.14), we need the second derivative of $\Sigma_{s,n}$ and $\Gamma_{s,n}$. From Eq. (A.11.6) we derive

$$
\frac{\partial^2 \Sigma_{s,n}}{\partial s_j \partial s_i} = H_s \frac{\partial^2 \psi_{s,n}}{\partial s_j \partial s_i} H_s^T + \Pi_{s,ij} + \Pi_{s,ij}^T \qquad \text{(A.11.15)}
$$

where

$$
\Pi_{s,ij} \equiv \frac{\partial H_s}{\partial s_j}\frac{\partial \psi_{s,n}}{\partial s_i} H_s^T + \frac{\partial H_s}{\partial s_i}\frac{\partial \psi_{s,n}}{\partial s_j} H_s^T + \frac{\partial H_s}{\partial s_j}\psi_{s,n}\frac{\partial H_s^T}{\partial s_i}
$$

and from Eq. (A.11.7)

$$
\frac{\partial^2 \psi_{s,n+1}}{\partial s_j \partial s_i} = \tilde{F}_{s,n}\frac{\partial^2 \psi_{s,n}}{\partial s_j \partial s_i}\tilde{F}_{s,n}^T + \tilde{\Omega}_{s_{ij}} + \tilde{\Omega}_{s_{ij}}^T \qquad \text{(A.11.16)}
$$

where

$$
\tilde{\Omega}_{s_{ij}} \equiv \frac{\partial \tilde{F}_{s,n}}{\partial s_j}\frac{\partial \psi_{s,n}}{\partial s_j}\tilde{F}_{s.n}^T + \frac{\partial}{\partial s_j}\Omega_{s_i}
$$

From Eq. (A.11.11) we obtain

$$
\frac{\partial^2 \Gamma_{s,n}}{\partial s_j \partial s_i} = \bar{H}_s \frac{\partial^2 \psi_{s,n}}{\partial s_j \partial s_i}\bar{H}_s^T + \nabla_{s,ij} + \nabla_{s,ij}^T \qquad \text{(A.11.17)}
$$

where

$$\nabla_{s,ij} \equiv \frac{\partial \bar{H}_s}{\partial s_j}\frac{\partial \tilde{\psi}_{s,n}}{\partial s_i}\bar{H}_s^T + \frac{\partial \bar{H}_s}{\partial s_i}\frac{\partial \tilde{\psi}_{s,n}}{\partial s_j}\bar{H}_s^T + \frac{\partial \bar{H}_s}{\partial s_i}\tilde{\psi}_{s,n}\frac{\partial \bar{H}_s^T}{\partial s_i}$$

and from Eq. (A.11.9)

$$\frac{\partial^2 \bar{\psi}_{s.n+1}}{\partial s_j \partial s_i} = \bar{F}_{s,n}\frac{\partial^2 \bar{\psi}_{s.n}}{\partial s_j \partial s_i}\bar{F}_{s,n}^T + \frac{\partial^2\left[\bar{G}_{s,n}\bar{Q}\bar{G}_{s,n}^T\right]}{\partial s_j \partial s_i} + \Delta_{s,ij} + \Delta_{s,ij}^T \tag{A.11.18}$$

where

$$\Delta_{s,ij} \equiv \frac{\partial \bar{F}_{s,n}}{\partial s_j}\frac{\partial \bar{\psi}_{s,n}}{\partial s_i}\bar{F}_{s,n}^T + \frac{\partial \bar{F}_{s,n}}{\partial s_i}\frac{\partial \bar{\psi}_{s,n}}{\partial s_j}\bar{F}_{s,n}^T + \frac{\partial \bar{F}_{s,n}}{\partial s_j}\bar{\psi}_{s,n}\frac{\partial \bar{F}_{s,n}^T}{\partial s_i} + \frac{\partial^2 \bar{F}_{s,n}}{\partial s_j \partial s_i}\bar{\psi}_{s,n}\bar{F}_{s,n}^T$$

From Eq. (A.11.13)

$$\frac{\partial^2 \bar{x}_{s,n+1}}{\partial s_j \partial s_i} = \bar{F}_{s,n}\frac{\partial^2 \bar{x}_{s,n}}{\partial s_j \partial s_i} + \frac{\partial \bar{F}_{s,n}}{\partial s_j}\frac{\partial \bar{x}_{s,n}}{\partial s_i} + \frac{\partial \bar{F}_{s,n}}{\partial s_i}\frac{\partial \bar{x}_{s,n}}{\partial s_j} + \frac{\partial^2 \bar{F}_{s,n}}{\partial s_j \partial s_i}\bar{x}_{s,n}$$

where

$$\frac{\partial^2 \tilde{\psi}_{s,n}}{\partial s_j \partial s_i} = \frac{\partial^2 \bar{\psi}_{s,n}}{\partial s_j \partial s_i} + \theta_{sij} + \theta_{sij}^T$$

and

$$\theta_{sij} \equiv \frac{\partial^2 \bar{x}_{s,n}}{\partial s_j \partial s_j}\bar{x}_{s,n}^T + \frac{\partial \bar{x}_{s,n}}{\partial s_j}\frac{\partial \bar{x}_{s,n}}{\partial s_i}$$

Note that some of the equations involved are commonly used in linear filtering and covariance analysis (Gelb, 1974) and in maximum likelihood identification (Gupta and Mehra, 1974; Sandell and Yared, 1978; Yared, 1979). Also note that the computation effort per step may be reduced (at the possible cost of reduced convergence rate) if instead of the exact Hessian matrix derived above, approximations are used. For instance, a Hessian approximation formula that employs the gradient values has been reported to perform quite effectively (Fletcher, 1979).

Chapter 12

Multi-Modal Aircraft Information-Guided Movement

12.1 Introduction

The problem of stochastic system control under large parameter variation or uncertainty is encountered in several applications and has received considerable attention in the past. A widely accepted approach to this problem is min-max optimization (Dem'yanov and Malozemov, 1974). A min-max solution for the deterministic controller design problem has been suggested (Salmon, 1968). Min-max filters and state estimators have been suggested and analyzed (Kassam and Lim, 1977; Poor, 1980; Poor and Looze; 1981). The min-max solution of the linear quadratic Gaussian control problem has been characterized for the case of fixed and known system parameters and unknown driving and measurement noise parameters (Looze, *et al.*, 1983).

Following an earlier study (Baram and Eidelman, 1985), we present a method for designing fixed-gain controllers and filters for systems with large parameter variation. This is the case encountered in aircraft control, where the dynamic characteristics are highly dependent on such variable parameters as altitude and Mach number (Athans *et al.*, 1977; Stein *et al.*, 1977). The approach is based on finding the min-max point of the Kullback information measure (Kullback, 1959) between the fixed-gain system and the optimal system at a given operating point. The design objective is to minimize the maximum possible difference between the optimal system at the actual operating point (i.e., the optimal adaptive system) and the selected fixed-gain system. The effectiveness of the proposed approach is illustrated by designing a longitudinal back-up control system for a given high-performance aircraft and evaluating its performance over a wide flight envelope. The resulting system compares favorably with a previously proposed technique over a wide range of flight conditions. Similarly, wide control regimes can be associated with uneven ground-based terrains, as well as inconsistent human movement due to neurological disorders. An understanding of the wide-envelope flight scenario can be highly instrumental in the design of a technological approach to biological movement disorders.

12.2 Design criteria

12.2.1 *General*

In the design problems under consideration, we have a stochastic system $M(p, s)$ depending on a parameter p whose values may vary over a set P and a parameter s whose value is to be selected from a set S in the design. Employing some optimal design criterion, to each parameter value p, there corresponds an optimal value $s(p)$ of s. It is desired, however, to find a single value s^0 that will be optimal in some sense for the entire parameter set P. The cost (or damage) of the resulting design at a given parameter value p is defined as the difference, in some measure, between the observations (or outputs) corresponding to $M(p, s)$ and $M[p, s(p)]$. Since the system under consideration is stochastic, it makes sense to measure the difference between the observations' probability density functions, corresponding to the two systems. A suitable measure is Kullback's information (Kullback, 1959), defined as

$$E_p^s \log\left[f_p^s(Y) / f_p^{s(p)}(Y)\right] \tag{12.1}$$

where Y is the observations set, $f_p^s(Y)$ and $f_p^{s(p)}(Y)$ are the probability densities of Y corresponding to $M[p, s(p)]$ and $M(p, s)$, respectively, and E^s denotes expectation with respect to $f_p^s(Y)$. This is a measure of the difference between the two densities when $f_p^s(Y)$ is the true density (the expectation is needed since Y is random).

In the present context, Eq. (12.1) measures the deviation of the output of the actual system $M(p,s)$ from that of the optimal one, $M\left[p, s\left(p\right)\right]$, at the operating point p. Reducing the cost implies that the performance of the actual system becomes closer to that of the optimal system at p. Since p varies over the set P, a reasonable objective Eq. (12.1) is to minimize the maximal cost, i.e.,

$$\min_{s\in S}\max_{p\in P} E_p^s \log\left[f_p^s\left(Y\right)/f_p^{s(p)}\left(Y\right)\right]$$

For the controller and the filter design problems considered in this chapter, the global min-max point would be obtained, in principle, by including in p both the controller and the filter parameters and performing the search simultaneously. ("Separation" of the two design problems does not appear to apply to the given criterion.) We have nevertheless chosen to solve the two problems separately for reasons of numerical implementation of the search algorithm and engineering interpretation of the results.

12.2.2 *Controller design for a linear system in finite time*

Consider a dynamic system in the form

$$\begin{aligned} x_{n+1} &= F_n\left(p,s\right)x_n + B_n\left(p,s\right)u_n + G_n\left(p\right)w_m \\ y_n &= H_n\left(p\right)x_n + v_n \end{aligned} \tag{12.2}$$

where x_0 is Gaussian with mean $m_0\left(p\right)$ and covariance $\psi_0\left(p\right)$, u_n is an external (non-feedback) deterministic input, w_n and v_n are zero-mean Gaussian white noise sequences with $E\left\{w_n w_k\right\} = Q_n\left(p\right)\delta_{nk}$, $E\left\{v_n v_k\right\} = R_n\left(p\right)\delta_{nk}$ and $W\left\{w_n v_k\right\} = T_n\delta_{nk}$. p is a variable parameter vector taking values in a set P and s is a vector of controller gains taking values in a set S. As shown in the sequel, any linear feedback of state or state estimate is accounted for in the matrix $F_n\left(p,s\right)$. We denote the above system by $M\left(p,s\right)$.

Employing some optimal control criterion, to each parameter value p, there corresponds an optimal gain value $s(p)$ and the corresponding system $M\left[p, s\left(p\right)\right]$. It is desired, however, to find a constant gain s^0 that will be optimal in some sense for the entire parameter set P. As in the case of optimal controller design at a given parameter value, the design may be aimed at optimizing the system's response to some input sequence or to an initial condition in finite time. Let us denote

$$I_N(p,s) = \frac{2}{N} E_p^s \log \frac{f_p^s(Y^N)}{f_p^{s(p)}(Y^N)}$$

where $Y^N = (y_1, \ldots, y_N)$. For the Gaussian case under consideration we have (Kullback, 1959; Hakes and Moore, 1976; Baram, 1976; Baram and Be'eri, 1981)

$$I_N(p,s) = \frac{1}{N} \sum_{n=1}^{N} \left[\log \frac{|\Sigma_n(p,s)|}{|\Sigma_n(p,s(p))|} + \mathrm{tr}\Sigma_n^{-1}(p,s)\Gamma_n(p,s) - k \right] \tag{12.3}$$

where k is the observations' dimension and

$$\Sigma_n(p,s) = E_p^s \left\{ \left[y_n - \hat{y}_{n|n-1}(p,s)\right] \times \left[y_n - \hat{y}_{n|n-1}(p,s)\right]^T \right\}$$

$$\Sigma_n(p,s(p)) = E_p^{s(p)} \Big(\{y_n - \hat{y}_{n|n-1}[p,s(p)]\} \times \{y_n - \hat{y}_{n|n-1}[p,s(p)]\}^T \Big)$$

$$\Gamma_n(p,s) = E_p^{s(p)} \left\{ \left[y_n - \hat{y}_{n|n-1}(p,s)\right] \times \left[y_n - \hat{y}_{n|n-1}(p,s)\right]^T \right\}$$

$$\hat{y}_{n|n-1}(p,s) = E_p^s \left\{ y_n \left| Y^{n-1} \right. \right\}$$

$$\hat{y}_{n|n-1}[p,s(p)] = E_p^{s(p)} \left\{ y_n \left| Y^{n-1} \right. \right\}$$

The equations for calculating the matrices $\Sigma_n(p,s)$ and $\Gamma_n(p,s)$ are provided in the Appendix. The fixed-gain controller is found from

$$s^0 = \arg \left\{ \min_{s \in S} \max_{p \in P} I_N(p,s) \right\}$$

12.2.3 *Regulator design*

Suppose that the system is time invariant, i.e.,

$$x_{n+1} = F(p,s)\,x_n + G(p)\,w_n$$
$$y_n = H(p)\,x_n + v_n$$

with

$$\mathrm{Cov}(w_n) = Q(p)\,; \quad \mathrm{Cov}(v_n) = R(p)$$

An attractive alternative, at least from the computation viewpoint, to finite-time design in this case is steady-state or regulator design. The following result is fundamental to our fixed-gain regulator design method.

Theorem 12.1. *Suppose that, for each $p \in P$ and each $s \in S$, the system $[F(p,s)\,,\, G(p)\,,\, H(p)]$ is controllable and observable and the controller $s(p)$ is stabilizing (i.e., $F[p, s(p)]$ and has all its eigenvalues inside the unit circle). Then the cost function $I_N(p,s)$ has a finite limit value, given by*

$$I(p,s) = \log\frac{|\Sigma(p,s)|}{|\Sigma(p,s(p))|} + \mathrm{tr}\Sigma^{-1}(p,s)\,\Gamma(p.s) - k \tag{12.4}$$

where $\Sigma(p,s), \Sigma[p,s(p)]$ and $\Gamma(p,s)$ are the limit values of $\Sigma_n(p,s)$, $\Sigma_n[p,s(p)]$ and $\Gamma_n(p,s)$, respectively. Δ

It should be noted that the controllability and observability conditions are structural requirements which are "generically" independent of the specific values of the parameters p and s. The condition that $s(p)$ is stabilizing is normally satisfied in pointwise optimal design. For instance, consider the system

$$\chi_{n+1} = F(p)\,\chi_n + D(p)\,z_n + \Omega(p)\,\omega_n$$
$$y_n = \Lambda(p)\,\chi_n + v_n \tag{12.5}$$

where ω_n is a noise process and z_n is a feedback process defined by

$$z_n = -s\hat{\chi}_n \tag{12.6}$$

where $\hat{\chi}_n$ is an estimate of the state. When $\hat{\chi}_n$ is obtained from the corresponding Kalman filter, Eq. (12.5) can be written as

$$x_{n+1} = F(p, s)\, x_n + G(p)\, w_n$$
$$y_n = H(p)\, x_n + v_n$$

where

$$x_n = \begin{bmatrix} \chi_n \\ \hat{\chi}_n \end{bmatrix}$$
$$F(p, s) = \begin{bmatrix} F(p) & -D(p)\, s \\ K(p)\, \Lambda(p) & F(p) - D(p)\, s - K(p)\, \Lambda(p) \end{bmatrix}$$
$$w_n = \begin{bmatrix} \omega_n \\ v_n \end{bmatrix}$$
$$G(p) = \begin{bmatrix} \Omega(p) & 0 \\ 0 & K(p) \end{bmatrix}$$
$$H(p) = \begin{bmatrix} \Lambda(p) & 0 \end{bmatrix}$$

It is well known that, by an appropriate choice of the gain matrix s, the eigenvalues of $F(p, s)$ can be placed inside the unit circle, provided that the pair $[F(\phi), D(p)]$ is stabilizable, e.g., (Athans, 1971). One such choice of s is obtained by means of

$$\min_{z_n} \sum_{n=1}^{\infty} x_n^T L x_n + x_n^T M z_n + z_n^T N z_n \tag{12.7}$$

where M and N are positive definite matrices and L is a positive semidefinite matrix, such that, for $C^T C = L$, the pair $[F(p), C]$ is observable. The operation in Eq. (12.7) yields z_n in the form of Eq. (12.6) with

$$s = s(p) = \left[D(p)^T \phi(p) + N\right]^{-1} \times [D(p)\, \phi(p)\, F(p) + M]$$

where $\phi(p)$ is obtained from the algebraic Ricatti equation

$$\phi(p) = K_1(p) + L - K_2(p)\, K_3(p)\, K_2^T(p)$$

where

$$K_1(p) = F^T(p)\,\phi(p)\,F(p)$$
$$K_2(p) = F(p)\,\phi(p)\,D(p) + M$$
$$K_3(p) = D(p)\,\phi(p)\,D(p) + N$$

Example 12.1. Consider the system in Eq. (12.5) with

$$F(p) = \begin{bmatrix} 0 & 1 \\ p_1 & p_2 \end{bmatrix}; \quad D(p) = \begin{bmatrix} 0 \\ 1 \end{bmatrix}$$
$$\Omega(p) = l; \quad H(p) = \begin{bmatrix} 1 & 0 \end{bmatrix}$$

and state feedback

$$z_n = -\begin{bmatrix} s_1 & s_2 \end{bmatrix} x_n$$

so that

$$F(p, s) = \begin{bmatrix} 0 & 1 \\ p_1 - s_1 & p_2 - s_2 \end{bmatrix}$$

It can be seen that the system $[F(p, s), G(p), H(p)]$ is controllable and observable and the pair $[F(p), D(p)]$ is stabilizable, independently of the values of $\mathrm{p}_1, p_2, s_1, s_2$. The conditions of Theorem 12.1 are thus satisfied and the cost function $J(p, s)$ has a finite limit value for all p and s. On the other hand, taking $F(p, s)$ as above but with

$$D(p) = \begin{bmatrix} 1 \\ 0 \end{bmatrix}; \; G(p) = \begin{bmatrix} 1 \\ 0 \end{bmatrix}; \; H(p) = \begin{bmatrix} 0 & 1 \end{bmatrix}$$

it can be seen that the system $[F(p, s), G(p), H(p)]$ is controllable and observable everywhere but on the line $p_1 = s_1$, and the pair $[F(p, s), D(p)]$ is controllable (hence, stabilizable) everywhere but at $p_1 = 0$. Since these points in the $P-S$ hyperplane produce indefinite values for the cost function $J(p, s)$, they may simply be avoided in the search procedure.

It should be noted that by virtue of the min-max criterion, the fixed-gain controller will be (output) stable everywhere in the parameter set, unless the parameter set is such that a globally stable solution does not exist. This follows from the fact that when the system is unstable, the matrix Γ_n and, consequently, the function $I_n(p, s)$ becomes infinite. This implies that the corresponding design parameters will not be selected in the minimization stage of the search procedure. The problem of global stability over parameter sets has been considered (Poor, 1980; Ackerman, 1980).

12.2.4 *Filter design*

Once a fixed-gain controller has been selected, the system is given by Eq. (12.2) with $s = s^0$. To each parameter p, there corresponds a linear stochastic model and an optimal estimator (Kalman filter) based on this model. It is now desired to find an estimator, independent of p (i.e., nonadaptive), that will be optimal in some sense for the entire parameter set P. Restricting the search to the set of Kalman filters defined on P, the min-max information criterion defined in Section 12.2.1 implies that the desired estimator is the Kalman filter corresponding to a parameter point $t \in P$, which is found from

$$\min_{t\in P} \max_{p\in P} E_p \log \frac{f_p(Y)}{f_t(Y)}$$

where $f_p(Y), f_t(Y)$ and E_p are the probability densities and the expectation associated with the corresponding parameters. (Note that if t is identified as the filter design parameter, then, using the notation of Section 12.2.1, we have $f_p^{t(p)} = f_p^p = f_p$, $f_p^t = f_t^t = f_t$ and $E_p^{t(p)} = E_p^p = E_p$.)

For an observation sequence $Y^N = (y_1, \ldots, y_N)$, let us denote

$$I_N(p, t) = \frac{2}{N} E_p \log \frac{f_p(Y^N)}{f_t(Y^N)}$$

then the filter design criterion in finite time may be written as

$$\min_{t\in P} \max_{p\in P} I_N(p, t) .$$

The calculation of $I_N(p, t)$ is similar to that of $I_N(p, s)$ in the controller design problem, replacing in the corresponding equations $F[p, s(p)], F(p, s), K(p, s), B[t, s(p)]$ and $B[p, s(p)]$ by $F(t, s^0), F(p, s^0), K(p, s^0), B(t, s^0)$ and $B(p, s^0)$, respectively, where s^0 is the selected controller gain vector, and replacing $\bar{H}(p)$ by $[H(t) - H(p)]$.

As in the case of controller design, steady-state estimator design offers a computational advantage over finite-time design. We have the following result.

Theorem 12.2. *Suppose that for each* $p \in P$*, the system* $\left[F\left(p, s^0\right), G(p), H(p)\right]$ *is stable, controllable, and observable. Then the cost function* $I_N(p,t)$ *has a finite limit value.* △

Proof. Substituting the corresponding matrices, the proof is identical to that of Theorem 12.1 □

It should be noted that an intermediate result of the proof is that under the condition of the theorem, the state estimation error

$$\psi_n(t,p) = E_t\left\{\left[x_n - x_{n|n-1}(p)\right]\left[x_n - x_{n|n-1}(p)\right]^T\right\}$$

has a finite limit value for any t and p. This means that although the nonadaptive filter is not optimal in the least-squares sense at any point but t, its mean-square error has a finite limit value. The fixed-gain steady-state estimator is obtained from

$$\min_{t\in P}\max_{p\in P} I(p,t)$$

where $I(p,t)$ is the limit value of $I_n(p,t)$.

12.3 Fixed-gain controller and filter design for aircraft

The proposed design procedure has been applied to the problem of finding fixed-gain values for the longitudinal control system of a given aircraft (the YF-16 CCV data has been used in the design). The linearized short-period equations are given by, e.g., (Etkin, 1972):

$$\begin{bmatrix}\dot{q}\\ \dot{\alpha}\\ \dot{w}\end{bmatrix} = \begin{bmatrix}a_{11}(p) & a_{12}(p) & a_{12}(p)\\ 1 & a_{22}(p) & a_{22}(p)\\ 0 & 0 & a_{33}(p)\end{bmatrix}\begin{bmatrix}q\\ \alpha\\ w\end{bmatrix} + \begin{bmatrix}d_1(p)\\ d_2(p)\\ 0\end{bmatrix}\delta_e + \begin{bmatrix}0\\ 0\\ g_3(p)\end{bmatrix}\xi \tag{12.8}$$

where q is the pitch rate, α is the angle of attack, w is a wind gust disturbance, δ_e is the elevator angle (no actuator dynamics are introduced here, for simplicity), and ξ is a zero-mean unity covariance white noise component of the wind disturbance. The coefficients $a_{ii}(p), d_i(p)$ and $g_3(p)$ are functions of the flight condition parameters $p = (h, M)$, where h is the altitude and M is the Mach number. Pitch rate and normal acceleration measurements are used for feedback purposes. The measurement equation is

$$y = \begin{bmatrix} q \\ a_{nt} \end{bmatrix} = \begin{bmatrix} 1 & 0 & 0 \\ 0 & a_{22}(p)\,U_0 & 0 \end{bmatrix} \begin{bmatrix} q \\ \alpha \\ w \end{bmatrix} + \begin{bmatrix} 0 \\ d_2(p)\,U_0 \end{bmatrix} \delta_e + v \tag{12.9}$$

where a_{nz} is the normal acceleration, U_0 is the forward velocity, and v is a zero-mean white noise process. Typical sensor accuracy and wind disturbance data can be found in the literature, e.g., (Athans *et al.*, 1977). The open-loop dynamical system was found to be unstable for most of the given flight conditions.

For digital control purposes, the equations have been discretized at 0.2 seconds sampling intervals. The resulting system has the form

$$\begin{aligned} x_{n+1} &= F(p)\,x_n + D(p)\delta_{e_n} + G(p)\xi_n \\ y_n &= H(p)\,x_n + v_n \end{aligned}$$

As a first step in the design procedure, an optimal controller is designed for each of the flight conditions. While any desired criterion could be used at this stage, we have chosen for illustration purposes the linear quadratic criterion (Stein *et al.*, 1977)

$$\min_{z_n} \lim_{N\to\infty} \frac{1}{N} \sum_{n=0}^{N} \left[\frac{q_n^2}{q_{\max}^2} + \frac{a_{nz_n}^2}{a_{nz_{\max}}^2} + \frac{\delta e_n^2}{\delta e_{\max}^2} \right]$$

It has been found (Athans *et al.*, 1977) that for all given p values, the pair $[F(p), D(p)]$ is stabilizable and the pair $[F(p), C(p)]$ is observable where

$$C(p) = \begin{bmatrix} \frac{U_0}{100g} & 0 & 0 \\ 0 & \frac{U_0^2 a_{22}(p)}{36g^2} & 0 \\ 0 & 0 & 0 \end{bmatrix}$$

Table 12.1. Optimal (linear quadratic) feedback gains for given flight conditions.

Flight condition number	Altitude (ft.)	Mach number	S_a	S_q	S_w
1	S.L.	0.6	−1.377	−2.048	−1.5242
2	15,000	0.6	−1.121	−1.101	−0.0413
3	30,000	0.6	−0.9603	−0.608	−0.0177
4	S.L.	0.8	−1.3916	−3.395	−0.0077
5	15,000	0.8	−1.197	−1.943	−0.0072
6	30,000	0.8	−1.079	−1.115	−0.0027
7	S.L.	0.9	−1.3175	−4.307	−0.0030
8	15,000	0.9	−1.1903	−2.575	0.0051
9	30,000	0.9	−1.1057	−1.525	0.0028
10	S.L.	0.95	−1.2048	−5.127	0.0261
11	15,000	0.95	−1.1086	−2.935	0.0456
12	30,000	0.95	−1.0575	−1.687	0.0225
13	S.L.	1.2	−1.6637	−6.782	0.1931
14	15,000	1.2	−1.4532	−4.044	0.2061
15	30,000	1.2	−1.3345	−2.344	0.1092
16	30,000	1.7	−1.7801	−2.698	0.1281
17	50,000	1.7	−1.6837	−1.383	0.0554

so that the resulting controller (regulator)

$$z_n = -s\,(p)\,\hat{x}_n$$

is stabilizing. In addition, the system $[F\,(p,s)\,,G\,(p)\,,H\,(p)]$ is controllable and observable for any s and p, so that the conditions of Theorems 12.1 and 12.2 are satisfied and the fixed-gain design criteria are well defined.

In searching for a fixed-gain controller, we have restricted ourselves to the finite set of optimal gain values $s\,(p)$ for the 17 flight conditions specified in Table 12.1. The fixed-gain controller design criterion now becomes

$$\min_j \max_j \{I\,(j,i)\,;\ i,j = 1,\ldots,17\}$$

The resulting fixed-gain controller gains are $s_a = -1.4532, S_q = -4.0437, s_w = 0.2061$. (These are the optimal controller gains corresponding to flight condition 14.) It should be noted that while a

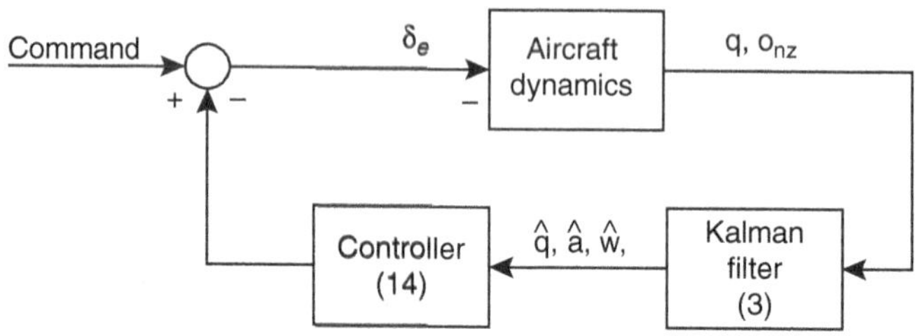

Figure 12.1. Resulting fixed-gain control system for aircraft.

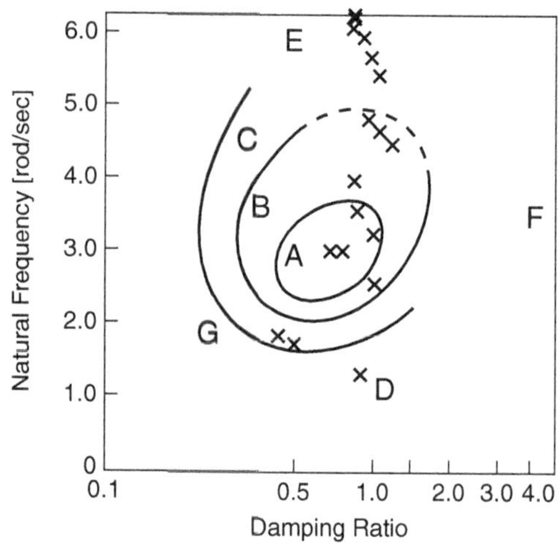

Figure 12.2. Performance evaluation – (A) satisfactory, (B) acceptable, (C) poor, (D) slow, (E) oversensitive, (F) sluggish, (G) unacceptable – of proposed fixed-gain design at given flight conditions by pilot rating contours (Baram and Eidelman, 1985). Performance evaluation of fixed-gain controllers obtained by previously proposed techniques (a) and (b) (Etkin, 1972, pp. 512–528).

full state feedback is considered here, as suggested in certain works, e.g., (Athans *et al.*, 1977), the wind gust variable feedback may be undesirable in certain situations. A partial state feedback of the variables α and q may then be found following a procedure similar to the one described above.

It remains to find a flight condition-independent state estimator. Substituting the fixed-controller gains into the dynamic models for each of the flight conditions, we get 17 "closed-loop" models. Employing the filter design criterion presented in Section 12.2.4 and restricting the search to the set of 17 optimal Kalman filters, the filter corresponding to flight condition 3 is found to be the overall (nonadaptive) optimal filter. The resulting control system is given schematically in Fig. 12.1, while performance evaluation characteristics at given flight conditions are presented in Fig. 12.2.

12.4 Conclusion

A method for designing fixed-gain controllers and filters for stochastic systems with large parameter variation has been presented. The approach is based on minimizing the maximal difference between the fixed gain system at an actual operating point and the optimal system corresponding to that point. The Kullback information measure defined on the systems' outputs is used as the measure of difference. The effectiveness of the proposed approach has been illustrated by designing a fixed-gain longitudinal control system for a high-performance aircraft. The resulting design has shown good performance qualities, although performance specifications have not been introduced directly into the design. This may be done by first designing each of the pointwise optimal controllers (at each of the flight conditions) so as to meet the specifications and then imposing the specifications as constraints on the min-max search procedure. In this chapter, we have described a fixed-gain control system for the entire flight regime. However, since the fixed-gain controller would normally be used as a back-up system in emergency situations, a small subset of the flight conditions corresponding to those situations may be considered in the design, so as to meet tighter performance requirements.

Finally, wide control regimes, as those associated with high performance aircraft, can be associated with uneven ground-based terrains, which may be encountered in human movement due to neurological disorders. An understanding of the wide-envelope flight scenario can be highly instrumental in the analysis of biological motion and the design of technological aids for the neurologically impaired, as addressed in the following chapters.

Appendix A.12.

The term $\Sigma_n(p,s)$ in Eq. (12.3) is obtained from

$$\Sigma_n(p,s) = H_n(p)\,\psi_n(p,s)\,H_n^T(p) + R_n(p) \tag{A.12.1}$$

where $\psi_n(p,s)$ is obtained from (Jazwinsky, 1970, p. 209)

$$\begin{aligned}\psi_{n+1}(p,s) = {} & F_n(p,s)\psi_n(p,s)\,F_n^T(p,s) + G_n(p)\,Q_n(p)\,G_n^T(p)\\ & - F_n(p,s)K_n(p,s)\left[H_n(p)\,\psi_n(p,s)\right.\\ & \left.+ T_n^T(p)\,G_{n-1}^T(p)\right]F_n^T(p,s)\end{aligned}$$

with

$$\begin{aligned}K_n(p,s) = {} & \left[\psi_n(p,s)\,H_n^T(p) + G_{n-1}(p)\,T_n(p)\right]\\ & \times\left[\Sigma_n(\pi p,s) + H_n(p)\,G_{n-1}(p)\,T_n(p)\right.\\ & \left.+ T_n^T(p)\,G_{n-1}^T(p)\,H_n^T(p)\right]\end{aligned}$$

($[\Sigma_n(p,s(p))]$ is obtained in a similar manner, replacing s by $s(p)$).

$\Gamma_n(p,s)$ in Eq. (12.3) is obtained from

$$\Gamma_n(p,s) = \bar{H}_n(p)\left[\bar{\psi}_n(p,s) + \bar{X}_n(p,s)\,\bar{X}_n^T(p,s)\right]\bar{H}_s^T(p) + R_n(p) \tag{A.12.2}$$

where $\bar{\psi}_n(p,s)$ is obtained from

$$\bar{\psi}_{n+1}(p,s) = \bar{F}_n(p,s)\,\bar{\psi}_n(p,s)\,\bar{F}_n^T(p,s) + \bar{G}_n(p,s)\,\bar{Q}_n(p)\,\bar{G}_n^T(p,s) \tag{A.12.3}$$

initialized at

$$\bar{\psi}_0(p,s) = \begin{bmatrix}\psi_0(p) & 0\\ 0 & 0\end{bmatrix}$$

and $\bar{X}_n(p,s)$ is obtained from

$$\bar{X}_{n+1}(p,s) = \bar{F}_n(p,s)\,\bar{X}_n(p,s) + \bar{B}_n(p,s)\,u_n \tag{A.12.4}$$

initialized at

$$\bar{X}_0 = \begin{bmatrix}m_0(p)\\ m_0(p)\end{bmatrix}$$

We have used the notation

$$\bar{F}_n(p,s) = \begin{bmatrix} F_n[p,s(p)] & 0 \\ F_n(p,s)\,K_n(p,s)\,H_n(p) & F_n(p,s)\,[I - K_n(p,s)\,H_n(p)] \end{bmatrix}$$

$$\bar{G}_n(p,s) = \begin{bmatrix} G_n(p) & 0 \\ 0 & F_n(p,s)\,K_n(p,s) \end{bmatrix}$$

$$\bar{Q}_n(p,s) = \begin{bmatrix} Q_n(p) & T_n \\ T_n & R_n(p) \end{bmatrix}$$

$$\bar{H}_n(p) = \begin{bmatrix} H_n(p) & -H_n(p) \end{bmatrix} \quad \text{and} \quad \bar{B}_n(p,s) = \begin{bmatrix} B_n[p,s(p)] \\ B_n(p,s) \end{bmatrix}$$

PART III

From Natural to Virtual Model Selection in Biological Motion Control

Chapter 13

Naturally Nonlinear Biological Motion Control

13.1 Introduction

In order to complement the linear dynamical models addressed in other chapters of this book, this chapter shows that natural movement has mathematically nonlinear elements. Specifically, we separately address two different scenarios of natural movement. The first scenario is an imminent collision with a visual object, in autonomous motion. Putting such occurrence in the context of subsequent chapters, we represent visual objects by tiles, and the event of collision by stumbling over a tile placed on the ground. Range, direction and time-to-contact can be extracted from the optical flow generated on the eye retina of a moving observer by the relative motion of rigid bodies (Koenderink and Doorn, 1975; Prazdny, 1981; Lee *et al.*, 1993). Such visual information may be used as a feedback signal to the movement control system for generating stumble-free movement (Baram 1999). The second scenario addressed in this chapter is visual object targeting and approach. Addressed in a control-theoretic context, the issue was put to the author by a NASA colleague seeking an answer to the following question: how can an aircraft pilot, wishing to land on airport runway, visually distinguish a real obstacle from a stain of oil. Recognizing the similarity of such situation to that of natural bird preying resulted in a simple mathematical calculation, leading to a rather simple answer (Baram, 1996).

13.2 Naturally nonlinear movement and stumble avoidance

Consider the observation geometry depicted in Fig. 13.1(a). The observer's vision is first assumed to follow a line intersecting the floor at an angle θ. The size of the field of view is represented by the viewing angle α. The tile arrangement is shown in Fig. 13.1(b). The observer's movement in the "N" direction creates a visual motion, called "optical flow", of projected tile edges on the observer's eye retina. The images of each of the tiles ahead of the observer expand with the forward motion. The motion of the projected edges of the center tile is represented by the arrows attached to the tile's edges. If the walker stumbles, the eye level reduces suddenly, while the forward motion continues. This results in a sudden change of the optical flow, as illustrated by Fig. 13.1(c). The actual visual flow depends on the actual motion of the eye, but, in general, the tile's image expansion rate during stumble increases drastically.

As will be shown mathematically, when the observer's velocity and eye level are constant, the net optical flow (NOF) of the center tile, representing the integral of the normal velocity along the projected tile boundary, grows approximately as the reciprocal of the forward velocity. As the observer moves forward, one center tile leaves the field of view and another enters. The walker's eyes fixate on the tile

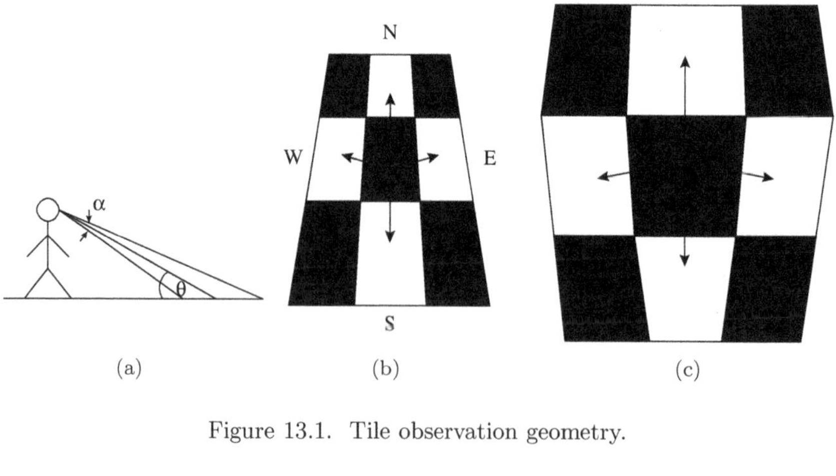

(a) (b) (c)

Figure 13.1. Tile observation geometry.

and visually tracks it, allowing the angle θ to grow with time. When θ reaches a certain threshold value, the observer's eyes make a fast movement to the next center tile (performing "saccade", see, e.g., (Hutton *et al.*, 1984)), and the process repeats itself. The resulting NOF signal is a steady "spike train", as depicted in Fig. 13.2(a). It serves as a reference signal, which may be learned by the observer in a few steps of steady walk, or from a standing position by eye motion, increasing the angle θ at a certain rate. A stumble will create an abrupt change in the actual spike train, as shown in Fig. 13.2(b), where the last spike is considerably greater than the

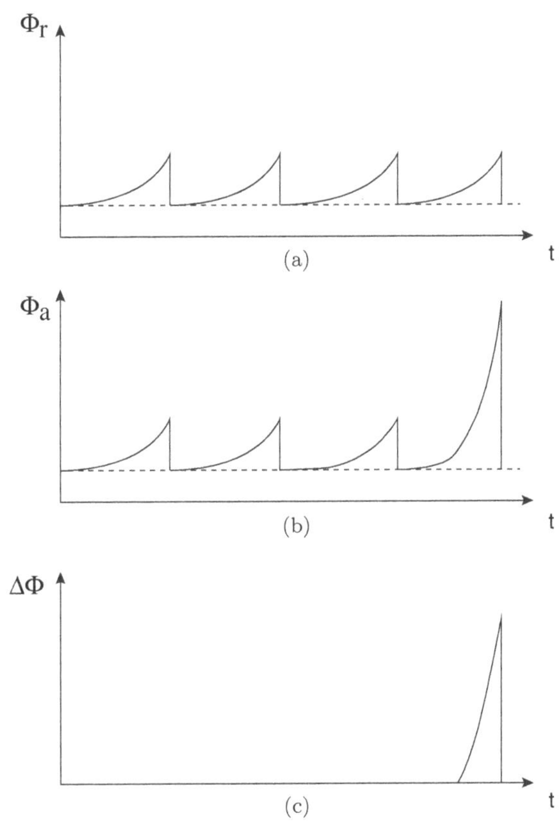

Figure 13.2. Spike trains ("saccades") of net optical flow generated by (a) a steady walk, (b) a stumble, and (c) the difference signal.

previous ones. The difference is a single, relatively large spike, as shown in Fig. 13.2(c). It constitutes a correction signal to the limbs, causing them to counteract a fall.

Clearly, even in healthy people, neither is the eye level constant nor are the visual tracking of tiles and the catch-up saccades so smooth as to create a uniform spike train as the one displayed in Fig. 13.2. Yet, the proposed idealization seems to represent the essence of the stabilizing visual effect provided by the tiles. In Parkinson's patients, the idealization seems even less realistic (Hocherman and Aharon-Peretz, 1994; Klockgether *et al.*, 1995; Vasilakos and Beuter, 1993; Visser, 1983; Hutton *et al.*, 1984; Zaccara *et al.*, 1992), although certain eye-tracking functions, such as prediction in smooth-pursuit, have been reported to remain intact (Fletcher and Aharpe, 1988). While the movement control impairment may cause severely irregular motion, the visual feedback signal, unsteady as it might be, provides, in essence, the same, and far more needed, stabilizing effect. It is like a cane which, held by a healthy walker can provide some unnecessary comfort, and, when held by a movement-impaired person, will provide, unsteady as it might be, essential support. The psychological effect is particularly interesting. It has been observed (Sacks, 1985) that the mere appearance of tiles in the field of view would cause a Parkinson's patient to start walking. This observation on the comforting effect of tiles seems to be supported by findings that more severely impaired patients display a higher anticipatory saccades activity than less severely impaired ones (Zaccara *et al.*, 1992). Relative tile motion generates such seemingly needed saccade activity. It has been suggested that pathological tremor present in Parkinson's patients interferes with their visual feedback (Vasilakos and Beuter, 1993). Expanding and contracting visual tile images may be generated by tremor and may serve in creating a feedback signal that activates the muscles, which, in turn, counteract these disruptions.

A mathematical analysis of the visual information generated by motion with respect to a single tile and the processing of this information can be based on the obstacle detection mechanism (Ringach and Baram, 1994). The sequential appearance of different

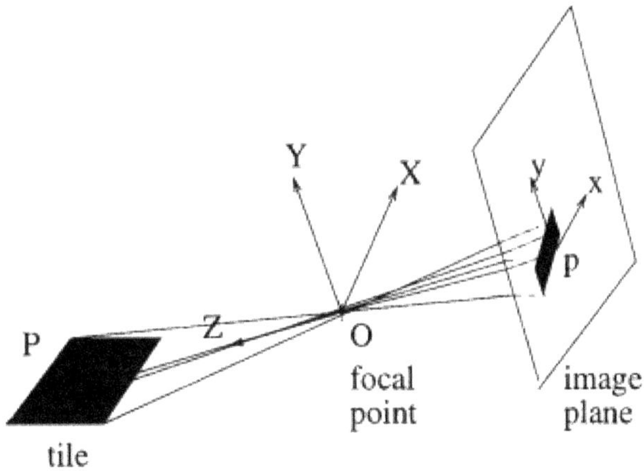

Figure 13.3. Tile projection geometry (Baram, 1999).

tiles in the field of view is handled by "saccades", as described above. Consequently, the corrective feedback signal is a sequence of "spikes", as depicted by Fig. 13.2(a), each representing the signal generated by a single tile. The tile projection geometry is shown in Fig. 13.3. A focal point O, representing the observer's location, is connected to a point P on a tile. An image plane, representing the retina, is placed at a certain distance from the focal point, perpendicularly to the line (OP), which intersects the image plane at point p. The points of the tile's boundary are connected to the focal point by straight lines, and the projected image of the tile is defined by the intersection of these lines with the image plane. Let (X, Y, Z) be a Cartesian coordinate system with origin at the focal point O and let (x, y) be a coordinate system in the image plane, with origin at point p. Let the relative velocity of the observer, located at O, with respect to the tile be given by $V = (V_X, V_Y, V_Z)$. Normalizing the velocities, $(\tilde{V}_X, \tilde{V}_Y, \tilde{V}_Z) = (V_X, V_Y, V_Z)/D$, where D is the distance between O and P, the optical divergence at a point p on the contour of the tile's projection on the image plane is given by (Ringach and Baram, 1994)

$$\Psi(p) \equiv \Delta \cdot v(p) = u_x + v_y = 2\tilde{V}_z + \left(\tilde{V}_x, \tilde{V}_y\right) \cdot (F_x, F_y) \qquad (13.1)$$

where $v(p) = (u, v)$ is the velocity of p and u_x and v_y are the respective partial derivatives in the x and y directions, and where F_x and F_y are the slopes of the tile with respect to the X and Y directions. It can be seen that in a steady walk forward at a constant velocity V_z with $V_x = V_y = 0$, the local divergence is inversely proportional to the time to contact with the obstacle $(t = D/V_z = 1/\Psi(p))$. When the observer looks at a point on the floor well ahead and walks steadily, both V_z and ΔF are nonzero. A stumble will cause abrupt changes in V_z and F_y, increasing both.

As has been observed, the local divergence represented by ψ(p) is a rather noisy signal when derived from real images and must be averaged over the projected image of an object (Koenderink and Doorn, 1975; Prazdny, 1981; Nelson and Aloimonos, 1989; Baram and Barniv, 1996; Baram *et al.*, 1997). For an untextured object, such as a tile, ψ(p) may be integrated along the contour of the tile's image (Ringach and Baram, 1994). Let R denote the projection of the tile on the image plane, $A(R)$ denote the area of the projected image, and ∂R denote the projected boundary of the tile. Further, let ds and $d\ell$ denote infinitesimal elements of R and ∂R, respectively. Then the average divergence of the points of R is

$$\Phi(R) = \frac{1}{A(R)} \int_{R/\partial R} \Psi(p)ds = \frac{1}{A(R)} \int_{\partial R} v_n(p)d\ell = \frac{1}{A(R)} \frac{dA(R)}{dt} \tag{13.2}$$

where $v_n(p)$ is the velocity normal to the edge at p. We see that $\Phi(R)$ can be calculated from the component of the velocity normal to the boundary and that it represents the relative rate of expansion of the tile's image. The averaging action in Eq. (13.2) suggests that the noise in $\Psi(p)$ will be largely eliminated and that $\Phi(R)$ will be a more accurate measure of the reciprocal time to contact with the tile. As before, it should be noted that in a steady walk, $\Phi(R)$ represents a certain rate of net optical flow through the tile's image, used as a reference signal in the feedback control system. Between saccades, the observer, moving forward at a constant velocity, follows each of the center tiles with his eyes. Consequently, the time to contact with the tile, as measured by the inverse of the average divergence, reduces approximately at a linear rate, equal to the forward velocity. This implies that the average divergence, or the net optical flow signal, increases between saccades at a rate which is inversely proportional to the forward velocity.

It has been shown (Ringach and Baram, 1994) that the calculation of $\Phi(R)$ is approximated by the final state of a network of locally connected array of cells, implementing a discrete diffusion process. Similar processes were previously proposed for performing other vision functions (Grossberg *et al.*, 1989; Grossberg, 1993). Assuming a two-dimensional grid arrangement of cells, as the one depicted in Fig. 13.4, the cell in position i at time $(k+1)\Delta t$, where Δt is some time interval, performs the calculation

$$u\,(i, k+1) = \frac{1}{|C(i)|} \sum_{p \in C(i)} u(p, k); \quad k = 0, \cdots \tag{13.3}$$

where $C(i)$ is the group of neighboring cells connected to the i-th cell and $|C(i)|$ is the number of cells in this group. This discrete

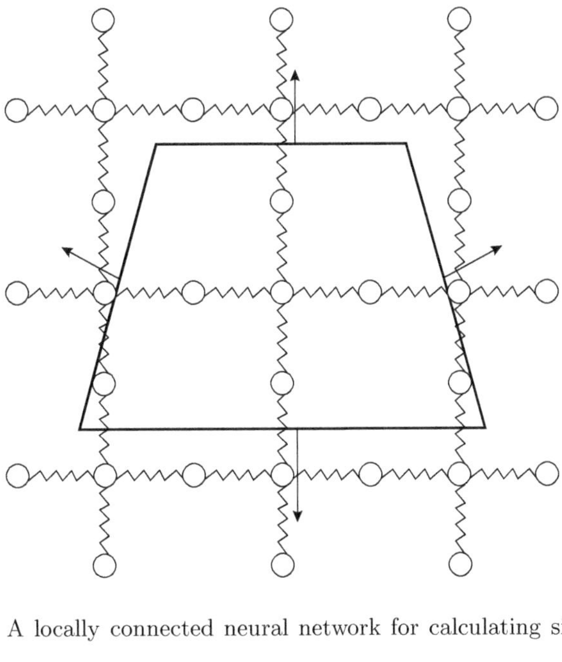

Figure 13.4. A locally connected neural network for calculating size change by diffusion.

diffusion process is initialized by zero everywhere but at the cells on the boundary of the image, which take the values (Marr and Hildreth, 1980)

$$v_n(i) = \frac{I_t(i)}{\sqrt{I_x^2(i) + I_y^2(i)}} \tag{13.4}$$

where I_t, I_x and I_y are the partial derivatives of the image light intensity with respect to t, x and y respectively. It has been shown that for a large number of nodes, the network is resistant to a high rate (nearly 50%) of connectivity failures (Ringach and Baram, 1992).

13.3 A bird's-eye view on the descent trajectory

When a bird descends towards a point on the ground for the purpose of landing or preying, it should be able to distinguish a three-dimensional object, constituting an obstacle or a target, from its two-dimensional background. The extraction of time to contact from optical expansion in birds was reported (Davies and Green, 1990, Lee *et al.*, 1993), and mechanisms for such calculations have been proposed (Nelson and Aloimonos, 1989; Ringach and Baram, 1994; Baram and Barniv, 1996). The simultaneous expansion of a background image can make target detection difficult. Following an earlier publication (Baram, 1996), we show next that the descent scenario facilitates a motion contrast between the optical images of a three-dimensional object and its two-dimensional background, making the distinction between them relatively easy.

Consider first a three-dimensional target at ground level and an observer descending along a certain line, as shown in Fig. 13.5. The observer's image of the target is represented by the viewing angle μ, which increases monotonically, as long as the observer's altitude is greater than the height of the target. When the target is replaced by a flat background patch, as shown in Fig. 13.6, the viewing angle of the background patch, θ, will exhibit a different mode of behavior. Denoting the width of the background patch a, its horizontal distance from the observer x, and the observer's altitude h, θ may be written

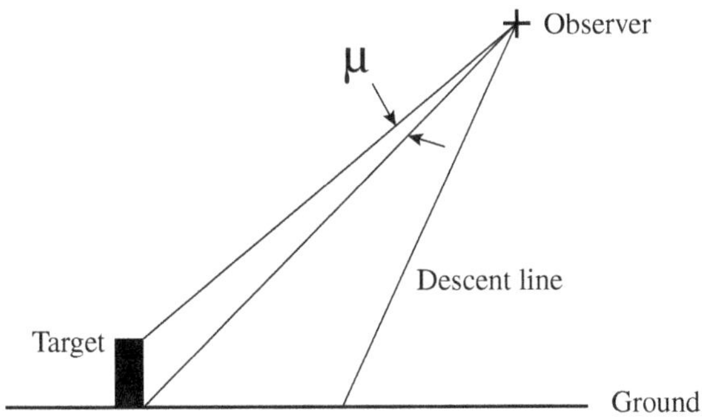

Figure 13.5. Target and observer geometry.

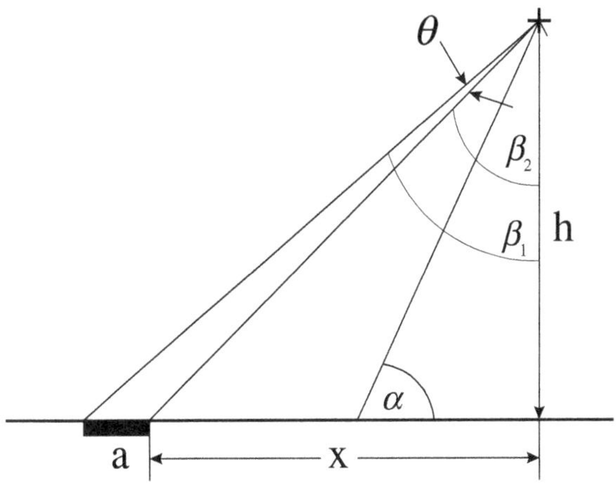

Figure 13.6. Flat background patch and observer geometry.

as the difference

$$\theta = \beta_1 - \beta_2 \tag{13.5}$$

where β_1 and β_2 are the angles between the vertical to the ground at the observer's location and the straight lines connecting the observer to the far and the near ends of the patch, respectively. We have

$$\tan \beta_1 = \frac{x+a}{h} \quad \text{and} \quad \tan \beta_2 = \frac{x}{h} \tag{13.6}$$

hence

$$\tan\theta = \frac{ah}{h^2 + x^2 + ax} \tag{13.7}$$

Denoting further, $b = x - h/\tan\alpha$, where α is the descent angle, we obtain

$$\frac{1}{\tan\theta} = \frac{1}{a}\left[\left(1 + \frac{1}{\tan^2\alpha}\right)h + \left(ab + b^2\right)\frac{1}{h} + \frac{a + 2b}{\tan\alpha}\right] \tag{13.8}$$

which relates θ to h along the line of descent, yielding a single maximum for θ at

$$h^* = \sqrt{\frac{b(a+b)}{1 + \tan^{-1}\alpha}} \tag{13.9}$$

The image of the background patch can be made to expand, contract, or remain stationary by choice of the descent angle α. Contraction or stationarity of the background's image will create a desirable contrast for the target's image, which keeps expanding in the vertical dimension. The greater α is, the faster the background contraction rate. However, in order to minimize the length of the descent trajectory, α should be as small as possible.

Selecting α locally so that the size of h along the descent trajectory, as represented by Eq. (13.6), provides the maximum viewing angle θ of the background patch, will guarantee a maximal view of the background without losing the contrast between the target and its background. It might be further noted that as in the case of obstacle detection discussed earlier, the image of the target is enhanced by an increase in the number of activated retinal and visual cortex cells, polarized positively by diffusion, as implied by image expansion, whereas the image of the background is progressively, albeit at minimal rate, silenced by negative polarization of the corresponding vision cells.

Taking then $h = h^*$ with $b = x - h/\tan\alpha$ and solving Eq. (13.4) for α, we obtain

$$\tan\alpha = \frac{h(1 + 2x)}{x^2 + ax - h^2} \tag{13.10}$$

The descent trajectory may be calculated for a background patch of a certain standard size, say $a = 1$ (equivalently, all lengths may be assumed, without loss of generality, to be given in units of a). It will then take the form

$$\frac{dh}{dx} = \frac{h(1+2x)}{x^2 + x - h^2} \tag{13.11}$$

which together with the initial condition (x_0, h_0) defines the descent trajectory.

Our analysis was based on the assumption that the local descent angle α is greater than $\tan^{-1}(h/x)$. This is clearly a necessary condition if the observer is to meet the target, with α changing monotonically. It follows from Eq. (13.11) that for the latter condition to hold at the initial point of the trajectory, we must have $h_0 \le x_0$. If this is not the case, the ratio x/h must be increased to 1 first. A bird may simply drop straight down to the point $h = x$, letting gravitation alone take effect. Then, spreading wings, it may follow the trajectory represented by Eq. (13.11).

Trajectories are shown in Fig. 13.7 for $h_0 = 50$ with $x_0 = 50$ (red), $x_0 = 100$ (blue) and $x_0 = 200$ (green). In the first case, the initial

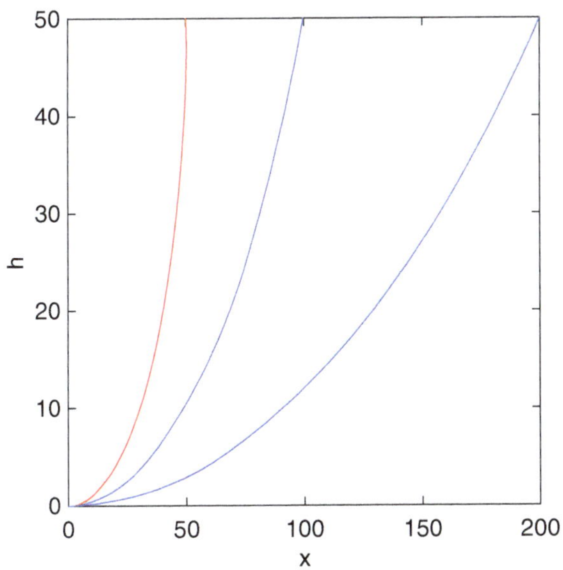

Figure 13.7. Descent trajectories for different initial conditions.

descent is implemented by a drop straight down. It can be seen that in all cases, a nice descent trajectory, ending at the target location, is attained.

It might be noted that the trajectory that maintains a constant viewing angle of the background patch in a global sense is the relevant part of the circle, which passes through the initial location of the observer and the two ends of the patch. There are, however, two major problems associated with flying along such a circular trajectory: it cannot be calculated locally from optical measurements alone, and it intersects the ground at a nonzero angle (which might have a disagreeable effect on any bird). In contrast, the proposed trajectory can be derived locally, by simply varying the descent angle until the viewing angle appears stationary. In terms of Fig. 13.7, the boundary of the image is replaced by that of the background patch viewed by the bird, and the diffusion process, causing changes in cell polarity, stops. Such local stationarity allows a gradual decay of the viewing angle. More importantly, the descent angle safely nullifies at zero altitude.

13.4 Discussion

Movement execution is perhaps the most fundamental task of the brain. Movement has been associated with enhanced neurogenesis, as well as learning and memory (van Praag *et al.*, 1999a; 1999b). Examining two specific natural movement tasks, one aimed at obstacle avoidance, the other aimed at object targeting, we have seen that, in contrast to the linear nature of the technologically driven control processes addressed in other parts of this book, the natural processes addressed in this chapter are largely nonlinear. The nonlinear effects do not reduce the accuracy of the resulting movement. Quite the contrary, they generate the desired outcome with mathematical precision. On the other hand, neurological impairments, as addressed in the following chapters, call for technological visual and auditory support, which are largely linear.

Chapter 14

Virtually Linear Biological Motion Control

14.1 Introduction

The previous chapter has addressed certain cortical processes which propagate sensory information so as to achieve natural movement objectives, specifically, object avoidance, targeting, and approach. The underlying mathematics of the control mechanisms were nonlinear. We now show that the inclusion of a technological mechanism, specifically, a portable virtual reality device, produces, by linear feedback control, improved gait entrainment in the neurologically impaired. Notably, model selection for improved movement, addressed mathematically in previous chapters (Chapters 8–12) is shown to be provided by cortical operations, specifically, vision, hearing, and motion control. We first mention a certain historical development, which, based on a scientifically sound early proposition, namely, visually-based feedback control, has later evolved in a fundamentally false direction, namely, open-loop control, before finding its way to highly effective, technologically sound feedback control.

A seminal study (Martin, 1967) and subsequent clinical tests (Bagley *et al.*, 1991; Morris *et al.*, 1994a; 1994b; Azulay *et al.*, 1999) have shown significant improvement in gait parameters of patients with Parkinson's disease (PD) walking over transverse lines drawn on the ground. Early attempts to generate such visual cues by portable virtual means have resulted in open-loop systems, imposing, by a constantly moving visual cue (Prothero, 1993; Weghorst *et al.*, 1994; Riess and Weghorst, 1995) or a geometrically patterned treadmill (Hanakawa *et al.*, 1999), a constant walking speed on the patient. A comparison of open-loop visual cueing by virtual means to transverse line markings on the ground (Griffin *et al.*, 2011) has found the first to have a marginal effect and the second to have a significant positive effect on gait parameters in PD patients. Subject-stationary visual cues (Lewis *et al.*, 2000) represent another unnatural imposition, where the image moves in the same direction as the subject, contrasting the opposite motion of an earth-stationary image. External open-loop auditory entrainment of gait, where a rhythmic sound is imposed on the patient in a metronome-like fashion, has been studied (Kritikos *et al.*, 1995; Thaut *et al.*, 1996; McIntosh *et al.*, 1997; Howe *et al.*, 2003; Nieuwboer *et al.*, 2007) and reasoned on neurobiological grounds (Thaut *et al.*, 2014). Yet, a need for constant vigilance and attention strategies to prevent reversion to impaired gait patterns caused by repetitive stimuli has been noted (Morris *et al.*, 1996). It has been further suggested that external rhythmic entrainment instigates dopamine reward (Miendlarzewska and Trost, 2013; Salimpoor *ct al.*, 2015), seemingly supporting open-loop intervention. However, open-loop control is known to be inherently unstable, with potentially disastrous consequences due to error accumulation (Kuo, 1981).

In sharp contrast, closed-loop feedback systems, when correctly designed, can regulate and stabilize otherwise unstable dynamics (Kuo, 1981). Analysis has shown that walking over earth-stationary visual markings constitutes a closed-loop feedback control system, which stabilizes and regulates gait (Baram, 1999), as shown in Chapter 12. Portable virtual realization of autonomous gait entrainment by sensory feedback, facilitated by the gradual miniaturization of sensing, computation, and display technologies, has led to the development of a closed-loop sensory feedback device for gait improvement in the neurologically impaired (Baram, 2004). While the effect created by the device is similar to that created by walking on a real floor with visual markings, the virtual earth-stationary visual markings are location-independent, as the patients are, in principle, free to go anywhere, making such entrainment completely autonomous.

14.2 Linear feedback control and autonomous cortical model selection

An examination of the natural sensory-motor control system underlying human locomotion with respect to the visual scenery reveals that it is the physical motion of the body that generates the visual cue and not the other way around (Baram, 1999). This observation is crucial to understanding the difference between external entrainment and autonomous entrainment of gait. The first represents a control system operating in open-loop, while the second represents a closed-loop feedback control system. The two control paradigms are illustrated in Fig. 14.1. In external, open-loop entrainment, a visual cue, moving at a constant speed, and/or a rhythmic auditory cue are generated artificially and fed through the eyes and/or the ears to the cortical motor lobe, which activates the limbs so as to respond to the sensory cues. On the other hand, in autonomous closed-loop entrainment, an artificial realization of the natural sensory-motor control system is a closed-loop feedback system, where the generation of the sensory

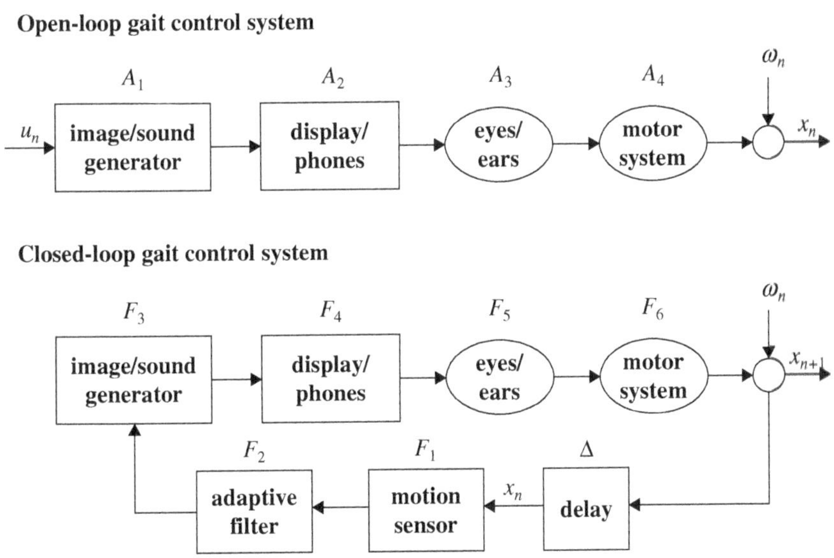

Figure 14.1. Autonomous (closed-loop) vs. external (open-loop) entrainment of gait.

feedback cue is controlled and regulated by sensory and movement control cortical elements. Here, the resulting motion of the visual cue and the rhythm of the auditory cue are matched to the motion of the body, which is normally matched to the steps executed by the legs. When there is no motion of the body, there is no sensory cue. The linear nature of both systems, open or closed loop, is characterized by the products of matrices, which may, for simplicity, materialize in scalar forms. The overall effects are products of linear operators.

The safety of autonomous gait entrainment is ratified by control theoretic considerations: closed-loop feedback systems, correctly designed, are, due to error correction, inherently stable, while open-loop, externally driven, systems are, due to error accumulation, inherently unstable; hence, unsafe (Kuo, 1981). As can be seen from Fig. 14.1, while error correction and noise elimination are made possible by the sensory feedback path of autonomous (closed-loop) entrainment, it cannot be implemented by the open-loop (no feedback) system of external entrainment.

The open-loop gait control system has the linear formula

$$x_n = Au_n + \omega_n$$

where

$$A = A_1 A_2 A_3 A_4$$

with $A_1, \ldots, A_4$ linear operators representing the corresponding components. While the first two, rectangular-shaped ("image/sound generator" and "display/phones") components are mechanically constructed, the last two, oval-shaped ("eyes/ears" and "motor system") components are cortical.

The closed-loop gait control system has the linear formula

$$x_{n+1} = Fx_n + \omega_n$$

where

$$F = \Delta F_1 F_2 F_3 F_4 F_5 F_6$$

with Δ the delay operator yielding

$$x_n = \Delta x_{n+1}$$

and $F_1, \ldots, F_6$ linear operators representing the corresponding components. Here, while the first five components are mechanically constructed, the last two, oval-shaped ("eyes/ears" and "motor system") components, are, again, cortical.

While the open-loop system is driven by a constant external input, the closed-loop system is driven by the patient's own forward motion. At the same time, while the effect of the external noise, ω_n, on the outcome cannot be corrected by the open-loop system, it is corrected by the feedback in the closed-loop system.

The autonomous model selection is controlled by the natural adjustment of the biological cortical components, namely, the "eyes/ears" and the "motor control" components. The biological construct of the two cortically controlled components have far-reaching capabilities of adapting to environmental, biological, and medical changes. Such adaptation corresponds to the notion of model selection, addressed mathematically in Chapters 8–12 of this book. The remaining components of the system are technologically contrived and constructed and, for the present considerations, are assumed to be unchanged during movement. Each box, albeit highly complex, may be approximately represented by a reduced linear system.

14.3 Virtual-natural motion-control in neurological impairment

Depicted in Fig. 14.2, a portable sensory feedback device, employing inertial sensors, an adaptive filter and a microprocessor contained in a belt-mounted cellphone-size box, is connected to a see-through micro-display, generating an earth-stationary visual feedback cue in response to gait entrainment (Baram, 2004). In addition to the visual feedback cue of glide-symmetric virtual tiles delivered by the display in coordination with the user's steps, the device also produces an auditory feedback cue in the form of a clicking sound delivered through earphones in response to every step taken by the patient. In contrast to open-loop, metronome-like devices, which attempt to impose a walking pace on the patient by a constant

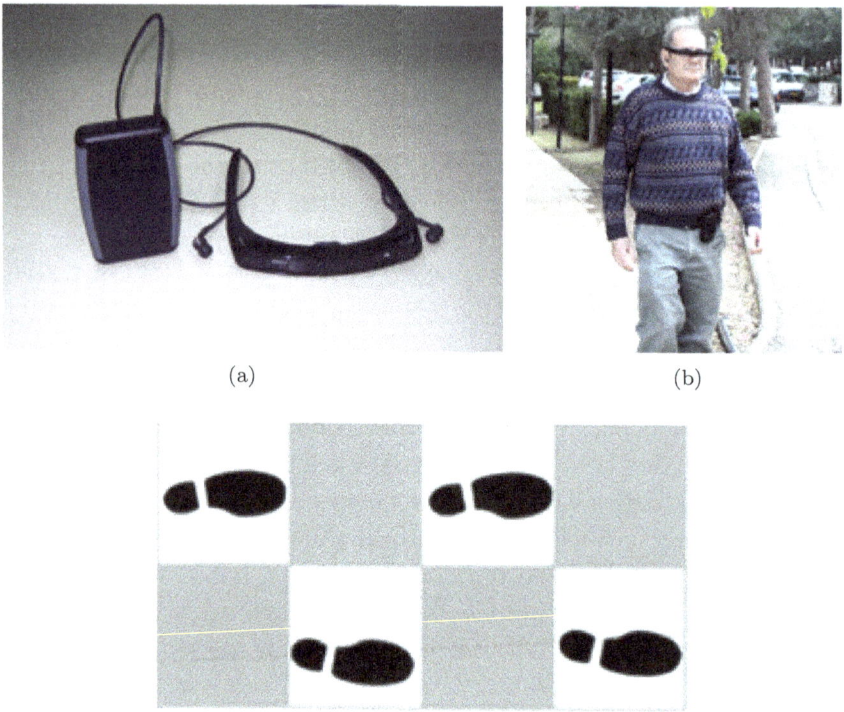

Figure 14.2. (a) Autonomous entrainment device. (b) Walking with the device. (c) Stepping on micro-displayed evenly spaced glide-symmetric virtual tiles.

auditory cue, the feedback device produces an auditory cue matched to the walking pattern. A balanced steady walk generates a rhythmic auditory cue. Any deviation from such a gait pattern will result in a deviation from the auditory rhythm and will be corrected by a change of the gait pattern in a feedback fashion. The head-mounted display and earphones bring the sensory feedback signals closer to the sensors – the eyes and the ears, making the sensory effect more pronounced, easier to follow. An open-loop capability, producing constant movement of the visual cue and a constant rhythmic auditory cue, was also added to an early version of the device for experimental comparison purposes. In contrast to the naturally nonlinear control mechanisms, mathematically and graphically presented in the previous chapter, the device virtually generates both visual and auditory linear feedback control. The linearity of the visual mode is manifested by step coordination with evenly spaced glide-symmetric virtual tiles, while the linearity of the auditory mode is manifested by rhythmic clicking sound generated by the contact of each of the steps with the corresponding virtual tiles.

14.4 Dopamine reward and reward-seeking

Before embarking on the presentation of clinical study results in subsequent chapters, it seems appropriate to pay attention to a certain internal cortical agent, which is believed to have a fundamental role in the well-being and physical performance of both healthy individuals and neurologically impaired patients. Such issues are believed to be associated with the presence, or lack thereof, of a neurotransmitter, called dopamine. A highly insightful study (Jenkinson and Brown, 2011) suggests that the frequency level of beta-range neuronal activity in the basal ganglia is modulated by the level of dopamine discharge at sites of cortical input, which are, in turn, modulated by salient internal and external cues. Moreover, it suggests that the level of the beta-range frequency provides a measure of the likelihood that a new voluntary action will be actuated. Put in our context, this insight suggests that autonomous entrainment of gait is more highly rewarded by dopamine discharge than rhythmic external cues. Yet, dopamine depletion in basal ganglia neurons of patients with PD suggests that it may be reward-seeking rather than actual dopamine reward, which comes into play in gait entrainment of patients with PD.

It has been further suggested that external open-loop entrainment instigates dopamine reward (Miendlarzewska and Trost, 2013; Salimpoor *et al.*, 2015), seemingly supporting open-loop intervention. However, as implied by clinical results, external entrainment, having some positive effects on some of the patients, also results in adverse effects, such as freezing of gait in patients with PD. These mixed results may be characterized as the "double-edged sword" of external entrainment.

It has been hypothesized that dopamine neurotransmission, highly evidenced in limbic areas of the basal ganglia (DiChiara *et al.*, 1994), elicits exhilaration or excitement in response to novel stimuli (Cloninger, 1987). The dopamine receptor gene D4DR has been linked to novelty-seeking behavior (Ebstein *et al.*, 1996, Benjamin *et al.*, 1996). In the context of cognition, novelty has been associated with "the ability to think and act independently" (The Free Dictionary, 2018). While it has been suggested that novelty-seeking, representing intellectual curiosity, aesthetic sensitivity, and risk taking, is low in dopamine-deficient patients with PD (Menza *et al.*, 1993), highly pronounced novelty-seeking behavior has been found in patients with PD, regardless of medication status (Djamshidian *et al.*, 2011). Dopamine reward has also been associated with vigilance (Aston-Jones *et al.*, 1994; Ikemoto, 2007), defined as "alert watchfulness" (The Free Dictionary, 2018).

While movement has been widely associated with reward, specifically, dopamine discharge by basal ganglia neurons, movement improvement in PD patients suggests the existence of a bypass mechanism. Prior reward learning, possibly facilitated by earlier normal preconditioning or medication, provides response incentive without actual dopamine signaling (Wassum *et al.*, 2011). In other words, learning to select responses that lead to reward results in reward-seeking behavior (Graef *et al.*, 2010; Moustafa, 2010). As novelty and vigilance have been widely recognized as key proponents of dopamine reward, the level of dopamine reward or reward-seeking should correspond to the levels of novelty and vigilance. While neither novelty, nor vigilance, have generally accepted measurable manifestations, both seem to be intuitively well-understood notions. In a cognitive perspective, novelty has been characterized as the ability to think and act independently (The Free Dictionary, 2018). In this respect, novelty does not seem to apply to external entrainment, which enforces a rhythmic auditory cue, or a constantly moving visual cue, on the patient. Indeed, it has been noted that the failure of external entrainment to produce satisfactory gait control in most patients is caused by the perpetual motion of the image, which, being unaffected by the patient, is eventually neglected (Morris *et al.*,

1996). In sharp contrast, autonomous entrainment, allowing the patient to decide if and when to take a step and at what stride length and speed of walk, is an embodiment of independent, free-willed, motion. Yet, vigilance appears to play a role in both external and autonomous entrainment, as both motivate a certain correspondence between step preparation and execution, on one hand, and the presentation of a sound or an image, on the other. The unequal balance between novelty and vigilance in the two cases suggests that while both external and autonomous entrainments present cases for dopamine reward, or reward-seeking, autonomous entrainment, instigating both novelty and vigilance, may present a much stronger case than external entrainment, which instigates vigilance only.

14.5 Discussion

The findings reviewed and compared in this chapter unambiguously show that autonomous entrainment of gait by visual and/or auditory feedback is not only decisively safer, but also significantly more effective than external entrainment. Specifically, it is the difference in retinal sensor polarity, generated by visual images, which entrains the motor system in a feedback fashion. While external gait entrainment is a "double-edged sword" producing adverse effects such as freezing in Parkinson's patients, autonomous entrainment comes without such adverse effects. While external entrainment represents an open-loop control system which is, due to error accumulation, inherently unstable, hence, unsafe for patient use, autonomous entrainment constitutes a closed-loop feedback control system which is, by error correction, inherently stable, hence, safe for patient use. Autonomous entrainment of gait by visual or auditory feedback produces significantly higher improvement in gait parameters than external entrainment. The checkerboard tiles geometry, being two-dimensional, produces in autonomous entrainment significantly higher levels of novelty and vigilance, resulting in significantly higher improvement in gait parameters than the one-dimensional transverse lines geometry. Yet, the level of gait improvement can vary significantly, not only with respect to the nature of the disorder, but also with respect to disorder severity and patient's own biological and biographical attributes, such as age, education, mental state, and cognitive abilities. Finally, it might be noted that hundreds of publications related to gait entrainment, while referring to autonomous (or closed-loop) entrainment, have, in fact, addressed external entrainment without making a distinction between external and autonomous entrainment. We find such presentations to be highly, although not necessarily intentionally, misleading. Following an early study presented in the following chapter, where we compared the gait improvement in Parkinson's patients employing the open and closed-loop versions of the device described above, all subsequent studies reported in the remainder of the book address closed-loop control alone.

PART IV

Gait Entrainment by Sensory Feedback in the Neurologically Impaired

Chapter 15

Walking on Virtual Tiles with Parkinson's Disease

15.1 Introduction

Gait impairment is one of the main complaints of advanced Parkinson's disease (PD) patients. Such patients often walk slowly with shuffling and dragging steps, reduced stride length, stooped posture, and diminished arm swing. Certain sensory stimuli are known to improve walking abilities in PD patients to a considerable extent. A pioneering study (Martin, 1967) has reported positive effects of earth-stationary visual cues, such as transverse lines, on gait in PD patients. While the positive effect of a walking stick as an aid to locomotion in patients with PD has been noted (Dunne *et al.*, 1987), the utility of visual markers has been further described by several investigators (Bagley *et al.*, 1991; Morris *et al.*, 1994a, 1994b, 1996; Azulay *et al.*, 1999). However, visual markers cannot be placed along each path the patient takes. Beneficial effects on gait in two PD patients using visual cues displayed by a portable head-mounted device have been reported (Prothero, 1993). The images were in constant perpetual motion, unrelated to the patient's own dynamics (the patient sees the same monotonous cue whether standing or walking, looking up or down). Clearly, this open-loop approach could not provide the safety needed by PD patients in movement. In contrast, the closed-loop earth-stationary virtual feedback cues generated by the patient's own motion (Baram, 1999) support the walking patient with complete stability and safety (see Chapter 12). Here, we first present the results of the seminal clinical study of patients with PD employing the closed-loop virtual feedback device (Baram *et al.*, 2002). We then present the results of subsequent studies, ratifying the original observations.

15.2 Seminal study of gait entrainment by virtual tiles in PD patients

The original device used in this study comprised a head-mounted 3-axis rotational accelerometer, a body-mounted 3-axis translational accelerometer, and a see-through head-mounted display, all connected to a wearable computer. It could be operated in two modes. In the open-loop mode, the virtual tiled floor performs perpetual motion towards the observer at a constant speed, unrelated to the patient's motion or lack thereof. In the adaptive closed-loop mode, the virtual floor appears to be fixed in space, as a real floor. When the patient moves forward, the virtual tiles appear to move in the opposite direction, again, as a real floor. In both cases, the virtual floor is superimposed on the real world, seen through the display. In contrast to the open-loop system, depicted in the upper part of Fig. 15.1, the closed-loop system, depicted in the lower part of Fig. 15.1, modifies the image of the virtual tiled floor produced by the image generator, in accordance with the patient's motions, measured by body-mounted motion sensor.

Fourteen PD patients of average age 68.2 ± 8.17 participated in the study. Mean disease duration was 9.36 ± 6.64 and mean clinical severity, according to HY staging (Hoehn and Yahr, 1967), was 3.04 ± 0.84. All patients had visual acuity 6/6 with correction when necessary. All tests were performed at the Movement Disorders Clinic, Rambam Medical Center, Haifa, Israel, around the same time in the morning, following 12 hours off medications. Before starting a test, each patient was made familiar with the device. Each test consisted of the patient walking a straight track of 10 meters four times. At the start of each test, the patient was verbally instructed to start walking. The time to complete the track and the number of steps for each path were recorded. In the first test (the reference test), the display was turned off. In the second test, the open-loop system was turned on, displaying a virtual tiled floor in perpetual motion towards the observer at the maximal speed level, which was still comfortable for the patient. The third test employed the adaptive closed-loop system. The order of the second and the third tests was then reversed and the results for each of the tests were averaged, in order to eliminate the effect of training from the comparison. The patients wore the device even when the display was turned off, so as to eliminate the burdening effect of the device from the comparison.

The test results are given in Table 15.1, where the last four columns show the percentage changes in the performance parameters obtained for the open-loop and closed-loop systems with respect to the reference test. It can be seen that, in all patients but one, performance was improved (higher speed, longer stride) significantly with respect to the reference test, when the closed-loop system was turned on. This measured improvement also represents the observed improvement in the quality of the steps. Patients who dragged their feet on the ground in the reference test raised them noticeably higher when the closed-loop system was turned on. The one patient who was not helped by the display at all, patient No. 11, had no walking impairment (as can be seen from his test parameters), and he had, by far, the best performance of all patients in the reference test. The last two rows in the table show the average performance of our patient group (excluding patient No. 11, who, as noted, had no walking impairment). It can be seen that, on average, the proposed closed-loop system improved performance.

It can be seen that, on average, the proposed closed-loop system improves performance by about 26% (speed) or 31% (stride length) with respect to the reference test. It should also be noted that the standard deviations of these results are rather high, which implies that the results should be evaluated mainly on an individual basis. Certain Parkinson patients would be helped by the proposed approach to a very significant degree (50%–100%), others would be helped to a lesser degree. Few, in particular those with no walking impairment, would not be helped at all. The average results for the open-loop system are about half as good as those for the closed-loop system. Furthermore, the standard deviations for the open-loop

Table 15.1 Performance results for PD patient employing closed-loop visual feedback (Yd: years of disease duration, HY: Hohen-Yahr staging, Speed: walking speed in meter/second, Stride: stride length in meters).

				Display off		Open-loop		Closed-loop		Open-loop Improvement %		Closed-loop Improvement %	
Patient	Age	Yd	HY	Speed	Stride	Speed	Stride	Speed	Stride	Speed	Stride	Speed	Stride
1	74	4	2	0.881	0.465	1.063	0.541	1.107	0.571	20.2	16.3	25.2	22.9
2	72	2	2	0.837	0.339	0.738	0.370	0.772	0.385	15.9	9.1	21.2	13.5
3	77	1	2	0.845	0.435	1.056	0.556	1.272	0.667	25.0	27.8	50.5	53.3
4	67	20	3	0.416	0.256	0.575	0.111	0.535	0.392	38.2	62.9	28.5	52.9
5	61	2	2.5	0.784	0.541	0.918	0.400	0.855	0.588	17.1	2.8	9.1	8.8
6	56	15	4	0.314	0.168	0.073	0.488	0.339	0.220	−77	−34	8.0	30.7
7	51	15	3	0.686	0.323	0.887	0.444	0.820	0.408	29.3	24.0	19.5	26.5
8	76	5	4	0.774	0.388	0.144	0.500	0.907	0.426	47.8	25.6	17.2	9.6
9	73	11	3	0.695	0.408	0.881	0.476	0.791	0.417	26.8	8.9	13.8	−2.1
10	57	3	2	0.620	0.385	0.907	0.400	0.887	0.488	46.2	30.0	43.0	26.8
11	73	10	3	1.038	0.555	0.855	0.488	0.888	0.500	−18	−14	−14	−9.9
12	82	20	4	0.508	0.339	0.601	0.140	0.590	0.385	18.3	18.0	16.0	13.5
13	59	10	4	0.773	0.435	0.864	0.434	0.895	0.488	11.8	12.2	15.9	12.2
14	57	13	4	4.405	0.154	0.243	0.145	0.668	0.531	−40	−9	65.5	12.8

system with respect to the means are drastically higher than those for the closed-loop system and for the reference state. This means that the open-loop system affects different patients in a drastically different way (indeed, for some of the patients the open-loop system gave slightly better results than the closed-loop one). The behaviors of patients numbers 6 and 14 are particularly noteworthy in this respect. Both these patients were severely bradykinetic (Berardelli *et al.*, 2001), unable to stand or start walking on their own. When the closed-loop display was turned on, both these patients were able to initiate and sustain a near-normal unaided walk. When the open-loop system (perpetual motion display) was turned on, both patients were able to initiate a walk but then experienced freezing episodes. It can be seen that for both patients, the performance parameters for the open-loop system are drastically lower than those for the closed-loop system and even those for the reference test. Some patients reported discomfort, dizziness, and nausea caused by the perpetual floor motion of the open-loop system. Most patients reported relative comfort with the self-activated, closed-loop adaptive system and indicated a clear preference for it over the open-loop system.

15.3 Recorded cortical sensory-motor activity

An early study (Hanakawa, 1999) employing blood-flow measurements has examined cortical activity in patients with PD assisted by treadmill-induced open-loop visual cue for gait entrainment. It has found that the right lateral pre-motor cortex, which is mainly regulated by cerebellar inputs, was activated to a greater extent in PD patients than in age-matched healthy individuals. On the other hand, the healthy individuals activated mainly the supplementary motor area (SMA), which was under-activated in PD patients.

A later study (Velu *et al.*, 2013), employing electroencephalography (EEG) recordings, has examined the cortical activity in patients with PD assisted by closed-loop visual feedback induced by a portable augmented-reality device, as described in the previous section. This case study has found a considerably higher increase in the beta frequency range of the connections between the occipital region and the motor region and between the occipital region and the parietal region of a responding patient with PD, compared to a non-responding PD patient, and to an age-matched healthy individual. Revisiting the results of that study, we find that they concur with the observations made in a recent book (Baram, 2021) on cortical polarity and singularity segregation. This is particularly noteworthy in view of the fact that polarity and singularity segregation have been argued in that book on predominantly mathematical grounds. The cortical recordings serve as direct and vivid evidence of not only the validity of cortical polarity and singularity segregation, but also their specific functional role in the directional connectivity between brain areas associated with the cortical functions of interest, specifically vision, navigation, and movement.

EEG (electroencephalography) is an electrophysiological monitoring method to record electrical activity of the brain. Employing a portable EEG device, cortical activity, relating neural firing rate to scalp-localized potential values, was recorded and compared in a cortical network composed of occipital (Oz), parietal (P4), and motor (Cz) channels (Velu *et al.*, 2013). The choice of these channels was based on anatomical findings from prior fMRI and PET experiments on paradoxical gait in PD (Hanakawa *et al.*, 1999). A patient with PD responding to visual autonomous entrainment (PDr), a non-responding patient with PD (PDnr), and a healthy age-matched individual (Control) participated.

Renormalized partial directed coherence (RPDC), representing directional connectivity between the cortical regions of interest, was used, as it provides a scale-free estimator, avoids normalization by outflows, and is not restrained by frequency dependent statistical significance (Schelter *et al.*, 2006). The experiment consisted of three walking stages: A (visual feedback turned off), B (visual feedback turned on), and C (visual feedback turned off).

The EEG spectra show color-coded RPDC differences between stages C and A in the delta (0–4 Hz), alpha (8–12 Hz), and beta (12–30 Hz) frequency bands corresponding to the connection between Oz and Cz (Fig. 15.1) and the connection between Oz and P4 (Fig. 15.2). Blue indicates a decrease, while red indicates an increase in the

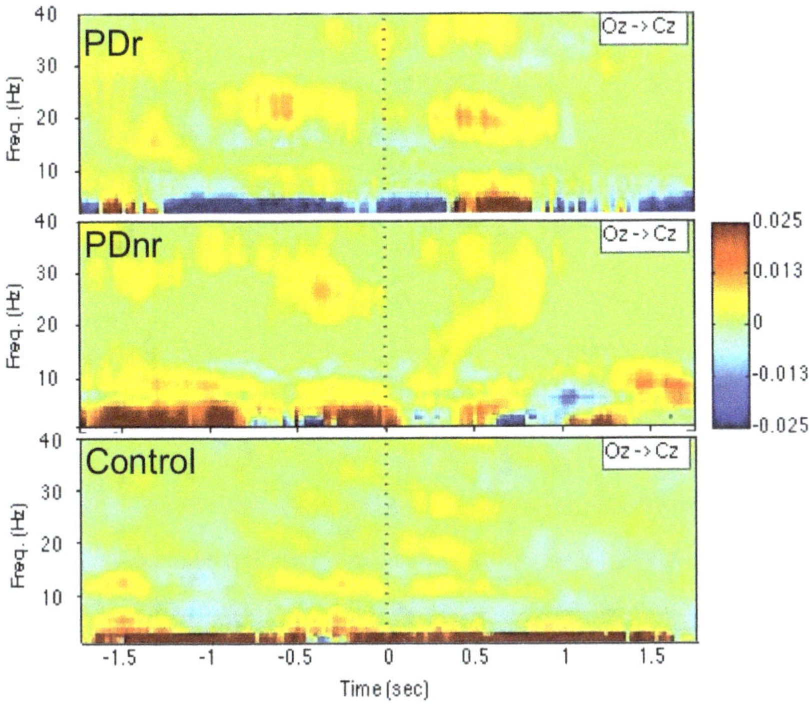

Figure 15.1. Color-coded renormalized partial directed coherence (RPDC), representing direction-specific connectivity from occipital region (Oz) to motor region (Cz). Blue indicates a decrease in RPDC while red indicates an increase in RPDC between Stage C and Stage A. An increased RPDC appears in the beta frequency range in PDr, reduced in PDnr, and only weakly noticable in Control. (Velu *et al.*, 2013).

differential RPDC (RPDC(C)-RPDC(A)). It can be seen that there is increased RPDC between Oz and Cz (Fig. 15.1) and between Oz and P4 (Fig. 15.2) in the beta range in PDr, reduced in PDnr and only weakly noticeable in Control. Similar results are seen in the difference in activity between stages B and A, but we only show the differences in RPDC between stages C and A to avoid any confound from neural responses to the presence of visual cues.

Putting the EEG results in the context of cortical polarity, red-colored RPDC represents positive polarity in the connectivity between the respective regions (Oz to Cz and Oz to P4), while

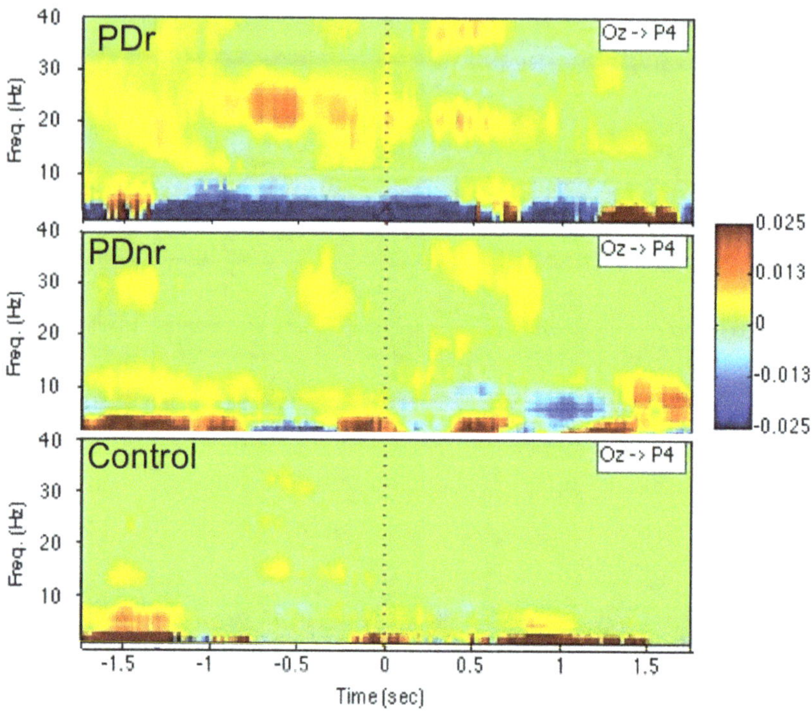

Figure 15.2. Color-coded renormalized partial directed coherence (RPDC), representing direction-specific connectivity from occipital region (Oz) to parietal region (Cz). Blue indicates a decrease in RPDC while red indicates an increase in RPDC between Stage C and Stage A. Occipital region (Oz) to motor region (Cz) flow shows increased RPDC in the beta frequency range in PDr, reduced in PDnr, and only weakly noticeable in Control (Velu *et al.*, 2013).

blue-colored RPDC represents negative polarity. It can be seen that in Fig. 15.1, showing Oz to Cz RPDC, there are positive polarity (red-colored) signatures in the beta frequency range (12–30 Hz) corresponding to PDr, considerably reduced in the beta range corresponding to PDnr, and almost non-existent in the beta range corresponding to Control. On the other hand, there is predominantly negative polarity in the delta frequency range corresponding to PDr, and predominantly positive polarity in the delta range corresponding to PDnr, considerably reduced in the delta range corresponding to Control. There is a very weakly noticeable activity in the alpha frequency range corresponding to all three participants. Essentially, similar results can be seen in Fig. 15.2, showing Oz to P4 RPDC. It might be noted that in the beta range for PDr, there are positive (red) RPDC signatures around both 0.5 seconds and -0.5 seconds in both Figs. 15.1 and 15.2. These signatures appear to correspond to two consecutive steps, about 1 second apart within the recorded walking sequence, centered at 0 seconds.

In the context of segregated singularity, the visible differences between the RPDC signatures corresponding to the different frequency zones, delta, alpha and beta, representing different connectivity sequences, indicate segregation between the modes of connectivity, hence, segregation of the corresponding cortical circuits. As shown in an earlier book (Baram, 2021), a difference in circuit connectivity would imply a difference in circuit firing dynamics, as, indeed, represented by the EEG activities corresponding to the three frequency domains. In closing, we note, again, that it was not our purpose here to point out the advantages of visual feedback for improvement of PD gait, which was done in the original study (Velu *et al.*, 2013), but, rather, to point out the presence of cortical polarization and singularity segregation, mathematically revealed and analyzed in this book, and ratified by the EEG recordings.

Case studies are highly valuable means for groundbreaking research. Their merits include: detecting novelties, generating hypotheses, applicability when other research design are not implementable, allowing emphasis on narrative aspect (in depth understanding), and educational values. The major limitations are: lack of ability to generalize, or to establish cause-effect relationship, danger of over interpretation, retrospective design, and distraction by focusing on the unusual. Their results, while understandably subject-dependent, can be rather conclusive. The results presented and analyzed in this chapter represent a positive case in point. Each of the subjects tested represents a widely recognized category. Among patients with PD, there are patients that respond to treatment and patients that do not. Controls are, normally, age-matched healthy individuals that do not show noticeable response to treatment. As the purpose of the study discussed here was to experimentally examine a theoretically founded hypothesis, regarding, essentially, a model, and, only remotely, a cure for a disease (which has been examined in clinical studies of large cohorts of patients before), a case study was an important first step. The conclusive results were highly ratifying of the theory tested here: cortical activity is driven by circuit polarization, and its dynamic singularities are highly segregated. Motivated readers and researchers are highly encouraged to conduct and expand such studies.

15.4 Follow-up clinical test results

Following the seminal study (Baram *et al.*, 2002), we have developed a more advanced portable miniature virtual reality device and tested it on a group of PD patients at the Carmel Medical Center in Haifa, Israel (Badarny *et al.*, 2014). The essential parameters examined were the same as those examined in the seminal study, with the additional testing of the residual effect (without the device) 15 minutes after the device-guided tests. The details of this follow-up study are described below.

Twenty randomly selected PD outpatients, 15 men and five women, average age 71.25 years, mean disease duration 5.28 years, and HY (Hoehn and Yahr, 1967) range 2.5–4 participated in the study. Participants were on their regular anti-Parkinson medication. Patients with considerable visual deficits not compensated by correction, ocular movement dysfunction, and gait disturbances due to neuromuscular diseases were excluded. Disease-related disability was assessed using the HY scale.

Assessment of patient ambulation was based on walking speed (meter/second) and stride length (meters), calculated directly from measurements of time, and number of steps needed to complete a fixed straight track of 10 meters.

Visual cues were generated by a belt-mounted unit, housing motion sensors and digital processing components and delivered by a micro-display, shown in Fig. 15.1. The display, attached to the eyeglasses frame, provides the patient with a virtual tiled floor in a checkerboard arrangement, responding dynamically to the patient's own motion, much like a real floor, fixed in space. The small size of the display and the see-through feature of its enclosure ensure that the field of view is effectively unobstructed.

All tests were performed at the Carmel Medical Center, Haifa, Israel, at about the same time in the afternoon. Examination comprised five stages, each consisting of the patient walking a straight track of 10 meters: baseline, online display off, online display on, and residual effects. At each stage, the length of time to complete the 10-meter track and the number of steps were recorded four times and averaged. At the start of each stage, the patient was instructed to "walk normally", so as to reduce motivational factor.

Stage 1: Walking without the device for measurements of baseline performance.

Stage 2: Walking with the device placed on the patient, but with the display turned off, in order to measure the pre-conditioning effect.

Stage 3: Walking with the display turned on in order to produce and measure the on-line effect of the virtual tiled floor. Patients were instructed to "walk on the virtual tiles", using their imagination in trying to "reach for the next tile" with each of their feet. No further restrictions, regarding stepping on tiles of specific color or not stepping on tile edges were imposed or suggested.

Stage 4: Walking without the device after a 15-minute break, in order to measure the short-term residual effect of the visual feedback cues.

Stage 5: Re-evaluation of baseline performance without the device one week after the first examination, in order to measure the long-term residual effect of the visual feedback cues.

The clinical test results are given in Table 15.2.

With the device mounted on, but the display turned off, the patients showed 1.75% improvement in walking speed and 2.61% improvement in stride length, on average, with respect to the baseline performance. These changes represent a negligible "placebo" effect. Turning the display on, 65% of the patients improved either their walking speed or stride length, or both, by over 10%, and 55% of the patients showed over 20% improvement. Following the

Table 15.2. Test results for patients with PD employing closed-loop visual feedback and its residual effects: walking speed (meter/second) and stride length (meters). DD: disease duration (years), HY: staging. "Display off" columns present walking speed and stride length results for patients wearing the device with the display turned off. "Display on" columns show on-line performance results with the display turned on. "15 min residual" columns present performance results after a 15-minute break, when patients walked freely with the device taken off.

Patient				Baseline		Display off		Display on		15 min residual	
Patient	Age/sex	DD	HY	Speed	Stride length	Speed	Stride length	Speed	Stride length	Speed	Stride length
1	80 M	10	3.5	0.782	0.500	0.752	0.488	0.790	0.677	0.867	0.533
2	75 F	3	2.5	0.669	0.408	0.628	0.400	0.627	0.400	0.747	0.435
3	77 M	3	3	0.848	0.476	0.816	0.465	0.954	0.541	1.059	0.588
4	61 M	1	3	1.066	0.526	1.071	0.526	1.310	0.625	1.147	0.571
5	79 F	19	3.5	0.680	0.408	0.686	0.400	0.618	0.444	0.735	0.426
6	83 M	5	2.5	0.675	0.426	0.746	0.444	0.606	0.454	0.815	0.454
7	67 M	10	3	1.229	0.645	1.368	0.694	1.426	0.769	1.530	0.741
8	60 M	1	3	1.100	0.606	0.980	0.555	1.373	0.769	1.238	0.667
9	53 F	3	3	1.074	0.625	1.161	0.645	1.100	0.588	1.314	0.645
10	62 M	5	3	1.291	0.645	1.316	0.667	1.356	0.667	1.517	0.741
11	68 M	9	3	0.980	0.513	1.021	0.526	1.358	0.714	1.285	0.588
12	80 M	4	2.5	0.829	0.500	0.933	0.526	1.151	0.588	1.033	0.571
13	78 F	5	2.5	0.602	0.417	0.694	0.435	0.752	0.454	0.849	0.500
14	77 M	5	2.5	0.818	0.513	0.946	0.556	1.063	0.645	1.104	0.588
15	73 M	1	3	0.539	0.308	0.591	0.339	0.808	0.476	0.873	0.500
16	70 M	3.5	3.5	0.370	0.200	0.232	0.150	0.380	0.208	0.375	0.227
17	61 F	7	2.5	0.422	0.323	0.460	0.333	0.439	0.333	0.440	0.345
18	74 M	5	3	0.154	0.376	0.133	0.427	0.294	0.343	0.282	0.374
19	72 M	5	3.5	0.027	0.028	0.022	0.026	0.020	0.025	0.033	0.030
20	75 M	1	4	0.389	0.263	0.486	0.357	0.525	0.408	0.479	0.357

15-minute break, 85% of the patients showed over 10% improvement and 65% showed over 20% improvement in either walking speed or stride length or both, without the device. The differences between residual and on-line performance can be attributed to burdening and learning effects associated with using the device for the first time, and are expected to diminish with further training. One week after participating in the first test (where the patient used the device for approximately 20 minutes), 11 of the patients returned for another "baseline" test. Of these patients, 64% showed over 10% improvement and 36% showed over 20% improvement in their baseline performance. Two of the patients reported that, a week later, they still walked on the virtual tiles in their minds.

In this study, we found that the on-line visual feedback-induced improvement in the walking abilities of patients on their regular medication schedule (56% of the patients showing over 20% improvement in either walking speed or stride length) is similar to that found in our previous study (Baram *et al.*, 2002) on patients taken off their medication for 12 hours (57% of the patients improving by over 20%). Furthermore, the present study shows that, following VR training, PD patients experience a sustained improvement in their walking abilities. This finding is in line with a recent case study, showing a sustained improvement in a single subject following one-month training with visual cues placed on the ground.

The fact that PD patients on regular medication schedule, trained with a visual feedback VR device for less than 20 minutes, have a persistent significant improvement in both stride length and walking speed with only a slight decline, is particularly encouraging. Some patients described it as a sensation of walking on the tiles in their mind. This interesting "learning process" can be supported by animal model studies suggesting that basal ganglia provide an internal non-specific cue to trigger movements and imply that PD involves a deficiency in this cueing (Fahn and Elton, 1989). However, PD patients typically rely upon external visual cues (Beck *et al.*, 1961). It is possible that such external cues perpetuate imprinting of internal cue, forming a "cue memory" that helps to maintain certain motor skills.

Since PD is a chronic progressive disease, we suggest that the visual feedback VR technique should be examined as a long-term procedure in addition to physical therapy, to maintain gait and postural performances in PD patients.

15.5 Discussion

Visual cues, long known to affect gait patterns, provide a channel for non-pharmacological therapeutic intervention in PD. In the present studies we have demonstrated that it is possible to use virtual reality cues, superimposed on the real world, to help PD patients control their gait. In particular, we have shown that the best effect can be achieved using a closed-loop display that responds to the patient's own motion. Specifically, our device displays a virtual tiled floor, which, as a real floor, appears to be fixed in space, moving with respect to the observer only when the observer does. In contrast, we have found that an open-loop, non-adaptive cue display causes adverse effects such as dizziness, nausea or even freezing. Predictive sensory cues, such as serial transverse lines, may trigger step transitions during walk. Furthermore, transverse lines may help a PD patient scale stride length, which is abnormally short otherwise, possibly due to underestimated muscle activity required to produce sufficient movement. We suggest that our closed-loop virtual display system produces the same effects as a real tiled floor. We further suggest that the failure of the open-loop virtual display system to produce satisfactory gait control in most patients is caused by the perpetual motion of the image, which, being unaffected by the patient, is eventually neglected. If, on the other hand, the patient attempts to keep attention on the perpetual floor motion, the relevant optokinetic system gets continuously stimulated, causing dizziness. In addition, the perpetual floor motion in the open-loop system does not forgive the patient for deviating from the set speed, resulting in a sense of confusion and being "out of sync". In contrast, the patient using the close-loop system has, by body motion, full control of the relative motion of the virtual floor, which, by bio-feedback, helps regulate gait. A previous study (Hanakawa *et al.*, 1999) explored the functional neuroanatomy which underlines visual guided gait in PD patients. It found that the right lateral pre-motor cortex, which is mainly regulated by cerebellar inputs, was activated to a greater extent in PD patients than in age-matched healthy individuals by visual transverse lines. On the other hand, the healthy individuals activated mainly the supplementary motor area (SMA), which was under-activated in PD patients. It appears, then, that visually enhanced gait employs different brain pathways in PD patients compared to healthy individuals, bypassing the impaired SMA function. The proposed augmented reality apparatus may serve as a mechanism for activating this alternative pathway. Our study reveals that the gait parameters, which are most sensitive to anti-Parkinson medication, and are improved by pallidotomy (brain surgery), namely, walking speed and stride length, can also be manipulated, to a similar extent and without some adverse effects, by a closed-loop display of virtual visual cues. The full extent of this approach within a comprehensive diagnosis and care program for movement disorder patients is proposed for further investigation. We note, in closing, that beyond the seminal study (Baram *et al.*, 2002), the studies reported above point out both short-term residual signatures following virtual visual patient training, which have also surfaced in our other studies (see subsequent Chapters) and EEG recordings indicating cortical activity matching gait entrainment.

Chapter 16

Auditory Feedback for Gait Improvement in Parkinson's Disease

16.1 Introduction

The effects of sensory cues on gait in Parkinson's disease (PD) patients have been noted as early as half a century ago (Martin, 1967) and studied extensively in following years (e.g., Dunne *et al.*, 1987; Bagley *et al.*, 1991; Morris *et al.*, 1994a; 1994b; Azulay *et al.*, 1999). However, the unique closed-loop visual feedback aspect of motion over the earth-stationary markings provided in that early study has been overlooked and largely misunderstood in early technological attempts to generate cueing by externally enforcing constant visual motion in an open-loop fashion (e.g., Prothero, 1993). As analytically demonstrated (Baram, 1999) and clinically tested in comparison to open-loop visual cueing (Baram *et al.*, 2002), the fuller extent of closed-loop visual feedback, imitating environmental effects, requires more advanced technological means involving motion sensors and real-time computation. Open-loop, metronome-like, rhythmic auditory cues, externally enforcing a gait pattern in a feedforward fashion, have been reported to have certain positive effects in PD (e.g., Kritikos *et al.*, 1995; Thaut *et al.*, 1996; Howe *et al.*, 2003; Nieuwboer *et al.*, 2007; Delval *et al.*, 2014). The combined effects of closed-loop visual and auditory feedback on gait in PD patients has been examined (Espay *et al.*, 2010). The isolated gait improvement results of closed-loop auditory feedback in PD patients (Baram *et al.*, 2016) are presented below.

16.2 Methods

16.2.1 *Subjects*

Sixteen randomly selected patients with PD, average age 69.94 ± 7.85 years, average disease duration 6.12 ± 4.69 years, and average HY staging 2.53 ± 0.56 participated. The study, performed at the Rambam and the Carmel medical centers in Haifa, Israel, was approved by the respective institutional Helsinki committees, and informed consent was obtained from each of the participants once the nature of the procedures had been fully explained. Participants were on their regular anti-Parkinsonian medication. Patients with considerable hearing deficit not compensated by correction, and gait disturbances due to neuromuscular diseases were excluded.

16.2.2 *Assessment of performance*

Assessment of patient performance was based on walking speed (meter/second) and stride length (meters), calculated directly from measurements of time and number of steps needed to complete a fixed straight track of 10 meters.

16.2.3 *Closed-loop auditory feedback apparatus*

A clicking sound is generated by a belt-mounted cellphone-size unit, housing motion sensors and digital processing components, in response to every step taken by patient, and delivered to patient by earphones. This allows the patient to maintain a steady balanced gait, controlling the auditory cue by changing the gait pattern and correcting any deviation from the intended auditory cue by correcting the gait pattern. It stands in contrast to metronome-like devices, which externally enforce a rhythmic cue without providing any feedback information on performance to the patient who, by "falling out of sync", may lose confidence and balance.

All tests were performed at about the same time in the morning. Examination comprised four stages, each consisting of the patient walking a straight track of 10 meters: baseline, display off, display on, and residual effects. At each stage, the length of time to complete the 10-meter track and the number of steps were recorded four times and averaged. At the start of each stage, the patient was instructed to attempt to produce a rhythmic auditory cue by maintaining steady balanced gait.

16.2.4 *Procedure*

Stage 1:

Walking without the device for measurements of baseline performance.

Stage 2:

Walking with the device placed on the patient, turned off, in order to measure the pre-conditioning effects.

Stage 3:

Walking with the device turned on in order to produce and measure the on-line effect of the device. Patients were instructed to walk so as to generate a rhythmic sound with their steps. No further restrictions were imposed or suggested.

Stage 4:

Walking without the device after a 15-minute break, in order to measure the short-term residual effect of the closed-loop auditory feedback cues.

16.3 Results

The clinical test results are given in Table 16.1.

We compared performance with device turned on to baseline performance, and to performance with the device turned off. We then compared short-term residual performance to baseline performance.

Putting the device on the patient with the device turned off showed an average increase of $2.02 \pm 5.44\%$ in the walking speed and an average decrease of $-0.6 \pm 1.99\%$ in the stride length with respect to baseline.

Table 16.1. Test results for patients with PD employing auditory feedback: walking speed (meter/second) and stride length (meters).

Patient				Baseline		Device off		Device on		15 min residual	
Patient	Age/sex	DD	HY	Speed	Stride length	Speed	Stride length	Speed	Stride length	Speed	Stride length
1 AW	72 M	9	3	0.782	0.425	0.797	0.417	0.835	0.435	0.903	0.465
2 YT	78 M	15	1.5	0.999	0.588	0.997	0.571	1.191	0.714	0.859	0.513
3 SM	65 M	16	2.5	1.174	0.541	1.358	0.555	1.241	0.541	1.323	0.588
4 BD	58 M	6	2	1.055	0.645	1.013	0.645	1.811	0.667	0.855	0.476
5 GT	62 M	5	2	1.350	0.645	1.316	0.667	1.338	0.667	1.375	0.667
6 AZ	70 M	2	2.5	0.782	0.455	0.785	0.455	0.891	0.513	0.925	0.519
7 BW	63 F	4	1.5	1.071	0.588	1.131	0.588	1.186	0.625	1.200	0.625
8 HM	78 F	5	2.5	0.737	0.444	0.785	0.476	0.825	0.488	0.828	0.476
9 LO	77 M	5	2.5	1.007	0.571	1.052	0.588	1.206	0.667	1.104	0.625
10 AF	73 M	1	3	0.822	0.454	0.842	0.465	0.927	0.500	0.921	0.500
11 YK	80 M	10	3.5	0.902	0.556	0.975	0.588	0.934	0.588	0.929	0.588
12 AS	77 M	3	3	0.877	0.500	0.893	0.500	0.914	0.526	0.941	0.526
13 SB	61 M	1	2.5	1.148	0.556	1.095	0.526	1.120	0.556	1.133	0.556
14 RA	67 M	10	3	1.305	0.645	1.285	0.645	1.340	0.714	1.475	0.667
15 ZB	83 M	5	2.5	0.823	0.454	0.777	0.435	0.837	0.465	0.806	0.444
16 YK	60 M	1	3	1.125	0.625	1.168	0.667	1.342	0.714	1.270	0.690

DD: disease duration (years), HY: staging. “Baseline” columns present walking speed and stride length results for patients before any use of the auditory feedback device. “Device off” columns present walking speed and stride length results for patients wearing the device with the device turned off. “Device on” columns show on-line performance results with the device turned on. “15 min residual” columns present performance results after a 15-minute break, when patients walked freely with the device taken off. Note: Patients 2 and 4 went into “off” in residual tests (by their own testimony).

Walking with the device mounted on and turned on, the patients showed, on average, $12.37 \pm 18.37\%$ improvement in walking speed and $4.30 \pm 3.64\%$ improvement in stride length with respect to baseline performance. The average improvement with respect to walking with the device mounted on but turned off was $10.72 \pm 19.53\%$ in walking speed and $6.77 \pm 6.57\%$ in stride length.

Noting that patients 2 and 4, by their own testimony, went into "off" during their short-term residual testing, and removing these patients' short-term residual performance results, we obtain $9.09 \pm 6.34\%$ improvement in walking speed and $6.52 \pm 4.36\%$ improvement in stride length compared to baseline performance.

Device-on performance can be seen to improve walking speed in 87.5% of the patients and stride length in 81.25% of the patients with respect to device-off performance. Average short-term residual performance can be seen to represent improved walking speed in 75% of the patients and improved stride length in 68.75% of the patients, compared to baseline performance. Taking patients 2 and 4 (who, as noted, went into "off" during their short-term residual testing) off the count here increases the improvement results in both walking speed and stride length to 85.71%.

We note that there is a high correlation between walking speed and stride length. Indeed, employing the results presented in Table 16.1, we find that the correlation coefficient between the two measures in baseline performance is 0.8672. The on-line (device-on) correlation coefficient between walking speed and stride length is 0.7714, while the residual correlation coefficient between the two measures is 0.8848. The strong positive correlation between walking speed and stride length, and vice versa, raises the question whether there is also a strong correlation between improvement in walking speed and improvement in stride length. Employing Table 16.1 again, we find that the correlation coefficient between the on-line improvement in walking speed and the on-line improvement in stride length compared to baseline performance is 0.1819. Similarly, the correlation coefficient between the on-line improvement in walking speed and the on-line improvement in stride length compared to device-off performance is 0.2182. On the other hand, the correlation coefficient between the residual improvement in walking speed and the residual improvement in stride length compared to the baseline performance is 0.9141. Clearly, residual improvement shows considerably higher correlation between walking speed and stride length. The reason for the large discrepancy between the on-line and the residual results in this respect is left for further research.

While we have reported changes in average walking speed and stride length, average changes in cadence could also be calculated by simply dividing average walking speed by average stride length. We note, however, that the interpretation of average change in cadence would be highly case-dependent. For instance, high average cadence may imply severe gait impairment in PD patients on one hand, or efficient locomotion in healthy individuals or slightly impaired patients on the other. A more informative measure would be the rate of change in cadence, indicative of gait impairment severity in PD, which, however, was not measured in this preliminary study due to lack of suitable equipment.

16.4 Discussion

Previously reported application of open-loop (metronome-like) auditory cueing has been found to increase or decrease walking speed in PD patients according to changes in the cue frequency above or below the preferred pace, respectively (Howe *et al.*, 2003). At the same time, stride length has not been found to be significantly affected by such changes or by open-loop auditory cueing altogether (Howe *et al.*, 2003; Delval *et al.*, 2014). At the preferred pace, open-loop auditory cueing has been reported to result in 4.2% average improvement in the PG score, combining walking speed and stride length, among a variety of posture and gait measures (Nieuwboer *et al.*, 2007). Our present study has found that closed-loop auditory feedback, releasing the patient from externally enforced cueing, results not only in relatively high average improvement in walking speed (12.4%) but also in non-negligible average improvement in stride length (4.3%). Notably, the participants in the present study were non-freezers, and, consequently, the effects of closed-loop auditory feedback on freezing have not been addressed here.

The present study has found a pronounced short-term residual effect of closed-loop auditory feedback on both walking speed (9.1%) and stride length (6.5%). In contrast, open-loop auditory cueing has been found to have no functional carry-over effects (Nieuwboer *et al.*, 2007). A study of combined visual and auditory closed-loop feedback (Espay *et al.*, 2010) has found long-term residual improvement in both walking speed and stride length. We suggest that the short-term residual effect of training with closed-loop auditory feedback, found in the present study, may present a gait improvement modality, which should be studied further, along with its long-term effects. While the discrepancy we found between the highly correlated walking speed and stride-length improvements in residual performance and the poor correlation between the improvements in these measures in on-line performance may be generally explained by the discrepancy between short-term and long-term learning and memory effects, we suggest this discrepancy for more detailed future research.

Finally, we note that, while a large majority of the patients tested have shown improvement in gait parameters as a result of closed-loop auditory feedback, the results also show relatively high deviation in the level of improvement among patients, which indicates that different patients may benefit differently from such intervention. Another interpretation may call for more extensive instructional patient preparation and training, which have not been attempted in the present study to a significant degree.

Chapter 17

At-Home Audio-Visual Training by Parkinson's Patients

17.1 Introduction

The key role played by sensory feedback in gait stabilization and control, i.e., where the feedback signal is generated by the patient's own motion, has been demonstrated by the gait improvements in patients with Parkinson's disease (PD) when they are provided with earth-stationary visual markings (Martin, 1967; Baram, 1999). At the same time, a clinical study addressing the role of visual feedback in maintaining postural balance in stance demonstrated that feedback increases postural stability in standing tasks, indicating a more prominent role for closed-loop (feedback) control over open-loop (feed-forward) control (Rougier, 1999). The development of a wearable virtual reality device driven by inertial sensors that delivers earth-stationary visual feedback cues has provided an opportunity to assess a closed-loop system of sensory (visual-auditory) feedback stimuli to aid dysfunctional gait in patients with PD. We sought to examine the effect of a wearable, closed-loop, visual-auditory cueing system on Parkinsonian gait in the "off" state, when the magnitude of gait impairment is expected to be largest. Preliminary clinical studies (Baram *et al.*, 2002; Baram and Miller, 2006) have demonstrated that the closed-loop feedback device significantly improves gait speed while in use and appears to induce a residual benefit for several days by dynamically delivering a tiled floor pattern adapted to the patient's own motion. It is believed that the optical flow provides a reassuring feedback effect, informing the patient of motion being performed and matching visual feedback from foot-tile matching to auditory feedback from clicking obtained for each step. These elements help produce a balanced rhythmic walk. In contrast to the open-loop strategies, this device also appears to produce and maintain a high degree of patient confidence, comfort, and safety (Baram *et al.*, 2002). To the best of our knowledge, the present study (Espay *et al.*, 2010), the details of which are specified below, is the first assessing the feasibility and effects of at-home training with a closed-loop sensory-feedback device in patients with PD.

17.2 Methods

17.2.1 *Subjects*

We recruited consecutive, consenting patients with PD who had off-state gait shuffling, festination, and/or freezing and a score > 1 on the Unified Parkinson Disease Rating Scale (UPDRS) III item 29, and who did not require continuous ambulatory aid from the Movement Disorders Center at the University of Cincinnati. Patients spent over 3 hours/day in the "off" state. We did not study patients with primary on-state freezing to maximize homogeneity of the target population. Patients had been on a stable dose of anti-Parkinson medication and required no medication adjustments during the course of the study. The following exclusion criteria were used: (1) prior functional neurosurgery; (2) musculoskeletal disorders, such as severe arthritis, knee surgery, hip surgery, or any other condition that the investigators determined would impair assessment of gait; (3) history of stroke; (4) cerebellar, vestibular, or sensory ataxia; and (5) Mini-Mental State Examination score smaller than 24.

17.2.2 *Visual-auditory walker*

The device is composed of a small measurement-computation unit attached to the patient's clothing, a head-mounted microdisplay, and earphones (Fig. 17.1). The measurement-computation unit is composed of a multiaxial accelerometer, a compass, and a microcontroller. The apparatus, operating in an adaptive closed-loop mode, displays a life-size virtual checkerboard-tiled floor superimposed on

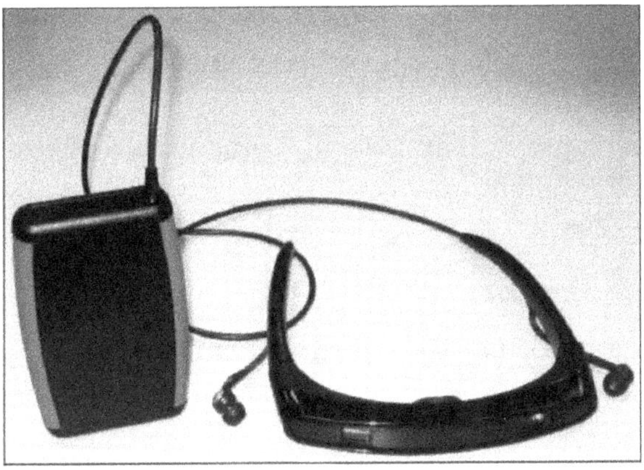

Figure 17.1. Virtual (augmented) reality device used in this study containing built-in LCD screen, which projects floor tiles when subjects are moving, and earphones that sound step-matched cue generated by connected sensor strapped at belt.

the real world with specialized see-through glasses. The closed-loop (or feedback) concept (as opposed to open-loop) implies that the speed of the cues is not externally set but, rather, is an outcome (and indeed a duplicate) of the walking speed of the user. The visual effect is the same as that created by walking over earth-stationary cues (i.e., a real tiled floor). Similarly, the rhythm of the auditory cue is determined by the rhythm of the steps, not vice versa. The user then regulates the gait pattern to create a constant optical flow and rhythmic auditory cue. The virtual (augmented) floor responds dynamically to the patient's own motion and "moves" toward him or her at the speed set by the user as measured by the accelerometer. The tiled floor acts as a moving visual display whose speed is generated in a natural feedback fashion by the patient's own motion, much like earth-stationary visual cues. The grid allows for patients to step on the tiles with long strides as they walk, though they do not enlarge or modify based on previous step length. A steady gait synchronizes the patient's own steps with the virtual tiles and the auditory cues, thus "rewarding" the patient for making the effort. Subjects can view real-world obstacles without difficulty. Subjects with corrective lenses are able to use the device without difficulty.

Additional auditory feedback from the patient's own steps is provided through earphones. Subjects hear the auditory cue produced by the steps through the earphones. The auditory feedback is continuous so long as patients are walking steadily, producing a rhythm they hear based on their gait pattern.

17.2.3 *Gait system*

The validated gait analysis system (GAITRite, CIR Systems, Inc; Havertown, Pennsylvania) used in this study consists of a 4-meter electronic walkway that contains six sensor pads encapsulated in a roll-up carpet to produce an active area 24 inches (61 cm) wide and 144 inches (366 cm) long (Webster *et al.*, 2005). In this arrangement, the active area is a grid (48 sensors by 288 sensors placed on 0.5-inch (1.27-cm) centers) totaling 13,824 sensors. The walkway is portable, can be laid over any flat surface, requires minimal setup and test time, and requires no placement of any devices on the patient. The application software controls the functionality of the walkway, processes the raw data into footfall patterns, and computes temporal (timing) and spatial (distance) gait parameters, including stride length, cadence, and velocity.

17.2.4 *Study evaluations*

Subjects were asked to walk at their usual pace on a defined 10-meter path encompassing the GAITRite walkway, beginning and ending at the same point under five conditions: (1) no sensory feedback (no device, baseline); (2) visual-auditory device positioned but deactivated (device "off"); (3) visual-auditory device positioned with visual-only feedback activated; (4) visual and auditory feedback activated, and; (5) again with no sensory feedback (device "off", immediate residual effect). Each condition was repeated three times, and the average of these was used for analysis. Due to the nature of the intervention, the order of the assessments was not randomized, and subject blinding was unattainable (signal noise in device would have worsened gait in any control group), but a comparison of conditions 1 (no device) and 2 (device "off") served to determine whether a placebo-like response could be elicited when the device was not yet providing visual-auditory feedback. Subjects were assessed at two time points: baseline (visit 1) and after 2 weeks of at-home use (visit 2). In between these in-clinic evaluations, subjects were instructed to use the visual-auditory device while walking for no less than two 30-minute periods a day, regardless of their medication state. We emphasized use during periods of expected off-related gait worsening. Subjects were not restricted in their choice of walking inside or outside and were encouraged to incorporate training with their regular activities. Since subjects were recruited on the basis of their off-related gait impairment, all study assessments took place during the "practically defined off period", that is, in the morning at least 12 hours since the last dose of any anti-Parkinson medication. We administered a standard motor examination with the UPDRS III (Fahn and Elton, 1989) and asked subjects to complete the self-administered Freezing of Gait Questionnaire (FOGQ, range 0–24; lower is better) (Giladi *et al.*, 2000) to ascertain their motor and gait-specific function.

17.2.5 *Data analysis*

We used the primary outcome measures (gait velocity, stride length, and cadence) to assess the effect of the device under various conditions. To accomplish this, we evaluated the effects of the various conditions (no device, device without feedback, visual-only feedback, visual-auditory feedback, and no device for immediate residual) on each outcome variable using repeated measures analysis of variance, followed by post hoc multiple comparisons with Bonferroni adjustment for each visit separately. The multiple comparisons were for each of the conditions versus baseline condition (condition 1). We examined the effect of training on each outcome using paired t-tests for each condition separately. Secondary outcome measures included changes in total FOGQ score, UPDRS III score, axial UPDRS III score, and freezing frequency between baseline and follow-up. We calculated the axial UPDRS III subscore by adding motor items 18–19 and 27–30, as described in a previous study (Espay *et al.*, 2005), to determine whether, in addition to gait, other axial motor features of

PD were affected by the device. We used paired t-tests to compare the secondary outcomes between the visits. As an exploratory analysis, we used multiple regression analysis to assess the effect of baseline UPDRS on the absolute change in velocity and stride length between condition 5 at baseline and follow-up, adjusting for disease duration. The significance criterion for the regression analysis was kept at 0.05. We considered p-values less than 5% significant.

17.3 Results

Fifteen patients were enrolled, but two did not feel comfortable using the glasses (because they were "clunky" or "embarrassing to use in public") and did not train at home as instructed. The patients (6 males, 73.3 ± 11.7 years old, disease duration 12.1 ± 4.2 years, levodopa equivalents $1{,}413.8 \pm 786.2$ mg) had a UPDRS III score ("off"-state during both assessments) of 28.1 ± 11.1 at the first assessment and 24.1 ± 9.1 at the second assessment (all data presented as mean $\pm$ standard deviation unless otherwise noted). Patients reported varying preferences of sensory feedback: four subjects favored the combination of auditory and visual feedback, three subjects preferred auditory stimulation alone, three visual feedback alone, and two neither (data missing for one subject). The gait parameters obtained for each condition with the gait analysis system are shown in Table 17.1.

17.3.1 *Clinical descriptions*

17.3.2 *Condition effect by visit*

At visit 1, a significant effect of the device on gait velocity was noted under various conditions ($p = 0.006$). The multiple pairwise comparisons revealed that condition 2 was not significantly different from condition 1 ($p = 0.39$). However, condition 3 was significantly different from condition 1 ($p = 0.008$). At visit 2, the effect of the device on gait velocity for various conditions was found to be statistically significant ($p = 0.03$). In post hoc pairwise comparisons, condition 5 was significantly different from condition 1 ($p = 0.01$), showing that a residual effect of the device on gait velocity existed.

Table 17.1. Gait parameters per condition (data presented as mean ± standard deviation).

Condition	Velocity (cm/s)	Stride length (cm)	Cadence (steps/min)
No device (baseline)			
Visit 1	64.2 ± 18.8	75.0 ± 15.4	104.4 ± 22.0
Visit 2	69.2 ± 22.8	81.1 ± 19.9	108.9 ± 23.3
%Δ	7.7 ± 14.8	8.1 ± 15.1	7.1 ± 23.0
Device without feedback (off)			
Visit 1	60.6 ± 18.3	73.9 ± 15.9	117.1 ± 27.3
Visit 2	71.5 ± 23.4	86.1 ± 23.7	114.8 ± 43.6
%Δ	18.7 ± 23.2	16.6 ± 19.4	−3.7 ± 15.6
Visual-only feedback			
Visit 1	56.4 ± 22.0	69.6 ± 20.5	105.3 ± 26.2
Visit 2	69.5 ± 26.8	80.4 ± 22.6	109.2 ± 24.5
%Δ	26.5 ± 29.1*	19.2 ± 28.4	5.9 ± 20.7
Visual-auditory feedback			
Visit 1	61.6 ± 20.1	74.3 ± 16.4	105.4 ± 32.0
Visit 2	72.6 ± 26.5	84.0 ± 18.5	110.6 ± 41.8
%Δ	17.1 ± 20.4†	13.6 ± 12.7†	3.3 ± 12.6
No device (immediate residual)			
Visit 1	64.5 ± 21.4	79.0 ± 20.3	107.4 ± 42.3
Visit 2	75.4 ± 21.5	88.8 ± 17.7	114.6 ± 25.8
%Δ	18.7 ± 14.4‡	14.9 ± 15.2‡	14.2 ± 35.0

*Significant difference between visits $p < 0.05$.
†Significant difference between visits $p < 0.01$.
‡Significant difference between visits $p < 0.001$.
Diff = Visit 2 − Visit 1, %Δ = percent change ((Diff/Visit2) × 100).

No significant changes in mean stride length were noted at different conditions on visit 1 ($p = 0.18$). However, a significant improvement in stride length was found at visit 2. Only condition 5 was found to be significant as compared with condition 1 ($p = 0.008$). Cadence was not significantly different under any condition for any visit. In summary, a condition effect on gait velocity as well as stride was found. The positive effects of the device on gait velocity and stride were mainly observed at the second visit between condition 5 and condition 1.

17.3.3 *Training effect for condition 1*

A trend for improvement in velocity (64.2 ± 18.8 cm/second to 69.2 ± 22.8 cm/second, $p = 0.05$) and stride length (75.0 ± 15.4 cm to 81.1 ± 19.9 cm, $p = 0.08$) was noted. Changes in cadence (104.4 ± 22.0 steps/minute to 108.9 ± 23.3 steps/minute) were not significant. Gait velocity improved by at least 10% in six subjects and by more than 20% in two. Stride length showed similarly graded responses (Fig. 17.2).

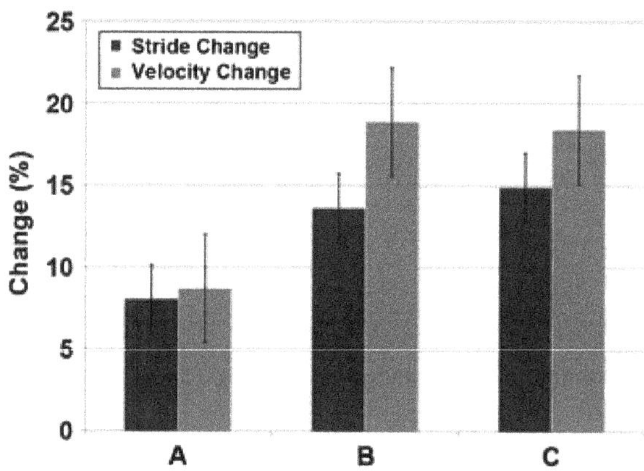

Figure 17.2. Percent change of velocity and stride length due to training effect (A = condition 1, day 1 vs. 14, no sensory feedback), combined device and training effects (B = condition 4, day 1 vs. 14, sensory feedback), and combined training and immediate residual effects (C = condition 5, day 1 vs. 14, no sensory feedback). Vertical bars denote standard error of the mean.

17.3.4 *Training effect for condition 2*

A significant effect was noted of training on gait velocity ($p = 0.02$) and stride ($p = 0.02$) but not on cadence ($p = 0.70$) for condition 2. This result indicates that improved training effects were obtained with the use of the device.

17.3.5 *Training effect for condition 3*

Under condition 3, significant improvement in gait velocity was noted ($p = 0.03$). Also, a trend toward improvement in stride length was found ($p = 0.06$). However, again, no change was noted in cadence for condition 3 ($p = 0.51$).

17.3.6 *Training effect for condition 4*

An improvement was found in gait velocity (61.6 ± 20.1 cm/second to 72.6 ± 26.5 cm/second, $p = 0.006$) and stride length (74.3 ± 16.4 cm to 84.0 ± 18.5 cm; $p = 0.004$). The improvement in cadence was not significant. Gait velocity improved by at least 20% in eight subjects and by more than 30% in five. Stride length improved by at least 10% in nine subjects and by more than 20% in five (Fig. 17.2).

17.3.7 *Training effect for condition 5*

Marked improvement was found in gait velocity (64.5 ± 21.4 cm/second to 75.4 ± 21.5 cm/second, $p < 0.001$) and stride length (79.0 ± 20.3 cm to 88.8 ± 17.7 cm; $p = 0.003$) but not cadence. Gait velocity improved by at least 10% in nine subjects and by more than 30% in three. Stride length improved by more than 10% in nine subjects and by more than 20% in five (Fig. 17.2).

17.3.8 *Training effect summary*

The results indicate that the training effect became more magnified under condition 4 and condition 5 for gait velocity as well as stride compared with condition 1. Also, a significant improvement in velocity and stride was found for condition 5 compared with condition 1, especially at visit 2. Thus, an effect of training and condition on velocity and stride exists.

17.3.9 *Unified Parkinson's Disease Rating Scale and Freezing of Gait Questionnaire*

Higher baseline UPDRS scores predicted residual velocity and stride length improvement ($p = 0.002$) adjusted for disease duration, indicating that patients with more advanced disease showed greater benefit. The FOGQ improved from 14.2 ± 1.9 to 12.4 ± 2.5 ($p = 0.02$). A trend toward decreasing frequency of freezing was noted from visit 1 (2.92 episodes/day) to visit 2 (2.54 episodes/day, $p = 0.09$).

The UPDRS III at visit 2 was 24.1 ± 9.1, lower than at visit 1 (28.1 ± 11.1) despite no changes in medications between the two off-state assessments ($p = 0.009$). A trend toward a reduction in the axial UPDRS scores was found at visit 2 ($p = 0.07$).

17.4 Discussion

Training with augmented reality visual-auditory feedback cues improved gait velocity by 18 to 20% in the immediate carryover state in a cohort of patients with PD who had substantial off-state gait impairment following a 2-week at-home period of self-training. Contrary to our expectations, the benefits were not observed within each visit, immediately upon device activation, or at the initial post-training evaluation before device activation, but as a robust immediate residual or carryover effect at the end of the last assessment. In fact, nearly 70% of the subjects showed residual improvement of at least 20% in velocity, stride length, or both (with almost 40% of the subjects improving by more than 30% in those measurements), changes that were significant despite high gait variability in response (as shown by large standard deviations). We speculate that the hardware required for the delivery of augmented sensory cues hindered the within-visit device-on response (by partially obstructing the visual field, adding discomfort with the belt sensor, etc.), which explains the greater benefits seen immediately after device removal compared with when the device was in use. This effect is expected to diminish with regular training and use of the device. Furthermore, improvement in the FOGQ, an instrument highly correlated with other measures of quality of life in PD (Moore *et al.*, 2007), suggests that the changes in gait were clinically significant. Patients with more advanced disease showed greater benefit after using the device, as greater baseline disability would provide the opportunity for a proportionally larger magnitude of benefit, whereas small baseline disability makes it harder for the intervention to show any beneficial effects. The reduction in at-home freezing severity noted in the FOGQ, although small in magnitude, is particularly relevant because of the pharmacoresistant of freezing and its association with an increased rate of falls (Gray and Hildebrand, 2000). A possible mechanism for the reduction in freezing, we speculate, is the generation of cortical reorganization or neuroplasticity, which may alter or circumvent the neuronal pathways responsible for this phenomenon. Finally, the improvement in velocity documented here was dependent on increases in stride length, at the relative expense of cadence. The lack of change in cadence is a reflection of the decrease in step frequency seen when slow to medium walking velocities are replaced with higher walking velocities resulting from increased stride length (Morris *et al.*, 1994a; 1994b).

This is the first study examining the effects of a virtual (or augmented) reality device for at-home use aimed at improving gait in patients with PD. A prior randomized crossover trial of an external cue training program provided at home by physiotherapists demonstrated small but significant improvements in the Posture and Gait Scale score, a composite of related UPDRS items (13–15 and 29–30) (Nieuwboer *et al.*, 2007). While that study was the first controlled trial that used a cueing device at home, it relied on an open-loop system with cueing parameters that needed separate adjustments to increase stride length and avoid festination and a training program led by a rehabilitation specialist. These features hinder its feasibility for wider application. Further, 67% of study subjects preferred auditory cueing over visual cueing (light flashes delivered through a light-emitting diode attached to a pair of glasses). Auditory cues may be less effective than visual cues (Cubo *et al.*, 2004).

The rehabilitation value of visual cueing, among other sensory feedback, was entered into the literature by an innovative neurologist (Martin, 1967). He suggested that the placement of visual cues perpendicular to the direction of gait spaced one step length apart was most effective in improving gait in patients with PD. Many single-session studies have confirmed the benefit of visual cues (high-contrast transverse floor lines) (Morris *et al.*, 1996; Lewis *et al.*, 2000; Jiang and Norman, 2006; Van Wegen *et al.*, 2006). However, these open-loop feedback systems may not have long-term effects

(Rochester *et al.*, 2007) unless a dedicated physical training program is established, as demonstrated in a patient trained to walk on floor cues reaching 120% of the uncued stride length over a 1-month period (Sidaway *et al.*, 2006). Closed-loop feedback systems do not require training program monitoring and may lead to long-term motor skills learning and enhancement of adaptive cerebral plasticity, particularly with the use of visual information on which the generation of motor plans is dependent in patients with PD (Helmich *et al.*, 2007). While an earlier study found closed-loop systems to be safer for patients (Baram *et al.*, 2002), selected patients may benefit from open-loop systems as well (Prothero, 1993). It is important to emphasize that adherence to a daily training program (minimum of 30 minutes of use twice per day) was critical to realizing the value of the closed-loop feedback system evaluated in this study, because immediate effects were not significant. The training effort needed at the initial visit was minimal, and patients encountered little difficulty in using this device at home without further guidance.

17.4.1 *Limitations of this study*

We must point out several limitations in our study. First, the absence of a control group was unavoidable given the device's lack of a "neutral mode" (e.g., white noise and nontiled visual feedback), which could have served as an "active" placebo. Also, due to safety reasons, we could not ethically justify a placebo arm in this at home, unsupervised gait rehabilitation type of intervention. As such, patients were unblinded to the intervention, and a potential order effect from the study conditions could have been present. The main outcome measures relied on electronic walkway rather than examiner-generated data to compensate for these methodological limitations. The lack of within-visit improvement in our patients, possibly related to subjects cautious approach to the study hardware, argues against a placebo response, although this does not rule out physiological test-retest variability in the gait of patients with PD. Second, although we requested and verbally confirmed compliance with the requested two 30-minute training sessions per day, we did not individually monitor device use at home. This precluded assessment of a dose-response effect and subjects preference for isolated visual or auditory feedback. The device was not uniformly accepted, and in two of the subjects, gait actually worsened while they were using it. In a similar vein, our small study did not have enough power to identify clinical variables associated with improvement versus deterioration. Similarly, the study design precluded our ability to ascertain the effect of single versus dual sensory cueing on gait, since subjects were instructed to use both visual and auditory cueing (though personal preference forced some to rely on only one or none). These issues, along with the decrease in gait performance with device use at baseline that was contrary to what was found in other device-based interventions, may cause concern regarding safe use of the device in a therapist-free intervention, compromised compliance, and increased variability, rendering the interpretation of our results preliminary. The high variability of effect, as well as its delayed occurrence, can very likely be attributed to the learning process associated with training with and using the device, which is highly variable and depends on patient's age and cognitive and mental capabilities. Finally, the off-state UPDRS scores were not matched at the two off-state assessments. The motor improvement at visit 2 was partly driven by a reduction in the axial scores, with which gait velocity has been found to correlate (Vieregge *et al.*, 1997) and was not explained by changes in medications or differences at the time of testing. This overall motor improvement may have acted as a confounder in the relationship between sensory cues and gait improvement. Prior studies have also shown nonpharmacological improvement in overall motor performance, such as home-based rehabilitation programs (Caglar *et al.*, 2005), and with other sensory stimulation, such as music (Pacchetti *et al.*, 1998). This is also highlighted by the discrepancy often found during routine clinic assessments whereby patients perform better (e.g., less freezing) than what they report to experience in their homes, possibly because of the heightened attention and expectation during the increased scrutiny of clinic visits. Indeed, the testing sessions at the office likely inflated our subjects' "true" off-state motor function and raised the functional baseline against which post-intervention gait parameters were compared.

17.4.2 *Future research*

Gait impairment is a major source of disability in patients with PD for which medical and surgical strategies offer relatively little. The noninvasive and effective sensory feedback device used here may prove to be a reasonable therapeutic option for some of these patients. Further studies should ascertain the features associated with responsiveness and lack thereof, the additive effect of combined sensory cueing versus visual- and auditory-only cueing, the length of the residual benefit after training (to determine the appropriate interval that may suffice to sustain a favorable gait performance in between uses of the device), the effect of this intervention on the less common but more intractable "on"-state freezing (Kompoliti *et al.*, 2000) and the benefit, if any, among postsubthalamic deep brain stimulation patients with treatment- and stimulation-refractory freezing of gait.

17.4.3 *Conclusions*

Devices using closed-loop sensory feedback through an at-home training program seem effective and desirable nonpharmacologic interventions to improve walking in patients with PD. The device helped people with PD by improving gait and decreasing freezing, which are expected to have an impact on functional mobility and quality of life. Further research is necessary to quantify duration of the residual benefit and applicability to other clinical situations in patients with PD.

Chapter 18

Virtual Reality Cues for Improvement of Gait in Patients with Multiple Sclerosis

18.1 Introduction

Gait disturbances in patients with multiple sclerosis (MS) may result from pyramidal syndrome (muscle weakness, spasticity), sensory (proprio-kinesthetic perception) disturbances, and general fatigue, as well as incoordination due to cerebellar ataxia. About 85% of patients with MS develop gait problems (Armutlu, *et al.*, 2001). Gait limitation results in mobility-related handicaps that can considerably limit the patients' activities of daily living, as well as their social interactions, and, therefore, have major impact on the patients, their families, and society (Moster and Kesselring, 2002). To date, most studies on gait rehabilitation, whether by means of physiotherapy, occupational therapy, or pharmacologic treatment, have focused on improvement of muscle strength and reduction of spasticity, such as pharmacologic approaches using Baclofen (Nilsen *et al.*, 2000; Orsnes *et al.*, 2000; Thoumie and Mevellec, 2002; Lord *et al.*, 1998). However, the issue of cerebellar ataxia has gained little scientific attention, resulting in relatively poor treatment options.

Recent research has focused on the search for gait assessment methods, in an attempt to lead to new approaches in intervention and rehabilitation. Typically, these studies consider gait in terms of time-distance measurements, i.e., speed and stride length, and observe their relation to muscle and joint movement analysis, muscular strength, and aerobic conditioning (Palliyath *et al.*, 1998; Benedetti, 1999; Morris *et al.*, 2002; Fauchard-Renard *et al.*, 2001; Mevellec *et al.*, 2003; Rodgers *et al.*, 1999). Patients with ataxia often use visual cues to maintain balance, which is the basis for the Romberg test. The application of virtual reality (VR) technology for the purpose of medical diagnosis and treatment has become widespread in the recent decades. Among its uses in neurologic rehabilitation is the development of motor assessment and training methods, using electromyographic, tactile (haptic), and visual cues (Ketalaer and De Wolf, 2000). A potential advantage of VR technology for the purpose of motor training is the capacity to provide the trainee with immediate performance feedback (Rizzo *et al.*, 2004).

It has been demonstrated that certain sensory stimuli improve the walking abilities of neurologic patients with movement disorders. In particular, moving visual cues, which generate a perception of motion, were found to have a beneficial effect on gait in patients with Parkinson disease (Martin, 1967; Azulay *et al.*, 1999). A study of the closed-loop effects of real-world visual cues on the regulation and stabilization of gait (Baram, 1999) has led to the development of a virtual reality apparatus (Baram, 2004). The closed-loop design employs a sensory feedback system to incorporate the patient's body motions in the display. The use of this apparatus was found effective for gait improvement in patients with PD above and beyond the effects of the open-loop version (Baram *et al.*, 2002). In the present study, we employed an advanced version of this device to study the effects of visual feedback cues on gait in patients with MS. Following an earlier publication, where additional details can be found (Baram and Miller, 2006), this chapter presents essential results demonstrating the utility and effectiveness of this approach in MS gait improvement.

18.2 Methods

18.2.1 *Subjects*

Sixteen randomly selected outpatients with MS, seven men and nine women, diagnosed according to the Poser criteria (Poser *et al.*, 1983) with gait disturbances predominantly due to cerebellar ataxia, participated in the study. Exclusion criteria were considerable visual deficit not compensated by correction, ocular movement dysfunction (e.g., diplopia, nystagmus), and gait disturbances due to pronounced muscle weakness, spasticity, sensory ataxia, or considerable general fatigue. Disease-related disability was assessed using the Expanded Disability Status Scale (EDSS) (Kurtzke, 1993). In addition, clinical assessment included Cerebellar Functional System Score (CFSS), which grades cerebellar function due to MS on a five-grade score, in which 0 = normal cerebellar function; 1 = abnormal signs without disability; 2 = mild ataxia; 3 = moderate ataxia; 4 = severe ataxia in all limbs or trunk; and 5 = unable to perform coordinated movement due to ataxia.

18.2.2 *Ambulation assessment*

Assessment of patient ambulation, conducted by an impartial independent investigator, was carried out using the ambulation measurements of walking speed (meter/second) and stride length (meters), calculated directly from measurements of time and number of steps needed to complete a fixed track. The control group consisted of 12 healthy individuals. The study was approved by the institutional Helsinki committee, and informed consent was obtained from each of the participants once the nature of the procedures had been fully explained. Clinical and personal data of patients and control groups are presented in Table E-1 (available on the *Neurology* website at www.neurology.org).

Visual cues were generated by a virtual reality apparatus (Baram, 2004): a closed-loop head-mounted device, shown in Fig. E-1. The display, attached to the eyeglasses frame, provides the patient with a virtual tiled floor in a checkerboard arrangement, responding dynamically to the patient's own motion, much like a real floor, fixed in space.

18.2.3 *Procedure*

All tests were performed at the Multiple Sclerosis Center, Carmel Medical Center, Haifa, Israel, at about the same time in the morning. Examination of each patient comprised four stages, each consisting of the patient walking a straight track of 10 meters: baseline, online display off, online display on, and residual effects.

Stage 1.

Baseline walking speed and stride length were measured first, without the device. The patient was verbally instructed to start walking. The time to complete the 10-meter track and the number of steps were recorded four times and averaged.

Stage 2.

The device was placed on the patient, but it was not turned on. Walking speed and stride length were again measured four times and averaged. The purpose of this stage was to produce a reference for the effect of the visual cue alone, without the burdening effect of the device.

Stage 3.

The display was turned on. The patient was instructed to walk on the virtual tiles along the 10-meter track. Walking speed and stride length were measured four times and averaged.

Stage 4.

The device was taken off the patient, who was given a 10-minute break. After the break, the patient was instructed to walk the 10-meter track without the device. Walking speed and stride length were recorded four times and averaged. The purpose of this stage was to measure the residual short-term therapeutic effect of the visual cues.

18.3 Results

18.3.1 *Clinical features, baseline measurements, and on-line improvement*

The test results for the patients and for the control subjects are summarized in Table E-2. Baseline walking speed and stride length of patients before using the device is given first. Walking speed and stride length of patients wearing the device, with the display turned off and on are given in the "on-line" columns, along with percentage changes between "display off" and "display on". Walking speed and stride length after the break, without the device, are given in the

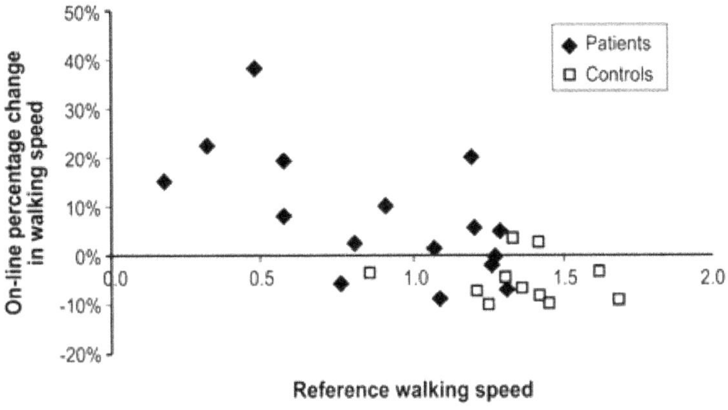

Figure 18.1. On-line percentage improvement in the walking speed of patients with MS and control subjects as a function of the baseline walking speed.

"residual effect" columns, along with percentage changes with respect to the baseline walking speed and stride length.

The percentage improvement in the on-line walking speed of the patients and of the controls with respect to the baseline walking speed is depicted for comparison in Fig. 18.1. Similarly, the residual improvement of the two groups is depicted in Fig. 18.2.

The medians of the baseline walking speed and stride length were 0.994 meter/second and 0.527 meters for the patients and 1.379 meter/second and 0.728 meters for the controls.

For the patient group, the on-line walking speed improved, on average, 13.46% in patients below the median and 1.47% in patients above the median. Using the baseline walking speed (BWS) value of 0.7 meter/second as a separation threshold, it was found that the on-line walking speed improved, on average, 20.32% in patients with BWS below the threshold and 1.63% in patients with BWS above the threshold. On-line average improvement in stride length was 9.78% in patients below the median and 5.74% in patients above the median. Using the baseline stride length (BSL) value of 0.45 meters as a separation threshold, it was found that the stride length improved, on average, 14.01% in patients with BSL below the threshold and 4.92% in patients with BSL above the threshold.

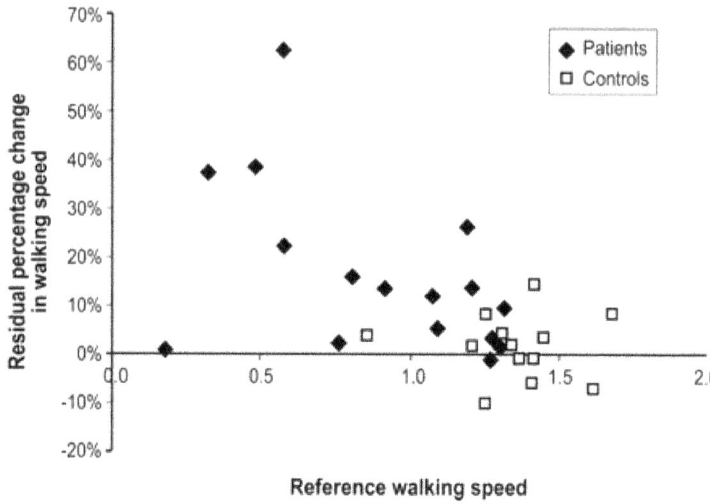

Figure 18.2. Residual (short term therapeutic) percentage improvement in the walking speed of patients with multiple sclerosis (MS) and of control subjects as a function of the baseline walking speed.

In contrast to the results for the patients, the results for the controls did not show any particular trend in relation to the BWS or stride length. Furthermore, using the device did not improve performance; quite the contrary, due to the burdening effect of wearing the device, it was found that the on-line walking speed changed, on average, by –5.3% in controls below the median and by –6.36% in controls above the median of the BWS. As in the case of the walking speed, there was an on-line decrease in stride length as a function of the BSL in the controls. The on-line average change in the stride length was also negative, –2.36% in controls below the median BSL and –1.08% in controls above the median. This makes the results for the patients even more noteworthy, since improved apparatus and prolonged training are expected to reduce the burdening effect, hence, further improve performance.

18.3.2 *Residual improvement*

The residual improvement in the walking speed (see Fig. 18.2) was, on average, 24.49% in patients below the median BWS and 9.09% in patients above the median. Average residual improvement in walking speed was 32.61% in patients with BWS below 0.7 meter/second and 9.59% in patients with higher BWS. In contrast, there was no notable residual improvement in the walking speed of the controls (2.49% and 2.36% changes in controls below and above the median BWS). The residual improvement in stride length of the patients was 16.32% in patients below the median BSL and 4.86% in patients above the median. Average residual improvement in stride length was 21.36% in patients with BSL below the 0.45 m threshold and 5.69% in patients with higher BSL. On the other hand, the controls did not show any residual improvement in the stride length (0.08% and –0.74% changes in controls below and above the median BSL).

During the tests, most of the patients reported a new sense of confidence and awareness. Following the tests, patients reported that an image of the tiles remains in their mind. One of the patients called a week after the test, saying that, when she walks, she is still walking on the tiles in her mind. Another patient reported in her subsequent visit that she walked better the whole week following the test.

18.4 Discussion

The advantage of Virtual Reality (VR) implementation in neurologic rehabilitation is the ability to control and monitor the delivery of stimuli (Rizzo *et al.*, 2004). Applying this principle, VR-assisted methods of rehabilitation have been devised to assess and treat both cognitive deficits and motor disorders. A case in point is the use of VR to provide visual feedback cues to improve gait control in patients with PD. It has been suggested that the series of transverse lines may help trigger step transitions during walk in patients with PD (Cohen *et al.*, 2011), and the stabilizing effect of such information has been shown (Baram, 1999). Consequently, the delivery of a tiled floor pattern, responding dynamically to the patient's own motion through a head-mounted VR device, has been suggested (Baram, 2004). When aided by the VR device, the gait quality of the patients with PD improved considerably (Baram *et al.*, 2002). The potential use of VR in the rehabilitation of patients with dysmetria, spasticity, and muscle weakness, using different stimuli, including haptics and EMG, has been proposed (Ketalaer and De Wolf, 2000). The results of the present study clearly indicate that it is possible to use VR cues, superimposed on the real world in a closed-loop fashion, to help patients with MS control their gait. The degree of improvement, both on-line and residual, was found to be inversely related to the BWS and stride length. These improvements in performance seem particularly noteworthy when compared to the results obtained for the controls, which showed no meaningful improvement. Above all, our findings regarding residual improvement are particularly encouraging: average short-term residual therapeutic improvement in walking speed was 24.49% in patients with BWS below the median and 9.09% in patients above the median.

The results of the present study support the potential role of VR-based strategies as rehabilitation modalities in MS, and substantiate their specific implementation in efforts to alleviate, improve, and restore mobility in patients with gait disturbances due to neurologic disorders in general.

Future studies may explore the implications of these results in larger cohorts of patients of various clinical subcategories (such as patients with above-median walking speed and below-median walking speed). An extended treatment course should be implemented, and its effects examined in terms of both the specific functional performance and long-term skill learning. Research that can link VR-enhanced rehabilitation to long-term learning would have added importance in MS, characterized by damage of coordinated interneuronal networks leading to associated dysfunction. Additionally, due to injury of Central Nervous System (CNS) reserves in MS, cerebral reorganization and plasticity may be limited (Cifelli and Matthews, 2002). VR-aided neurorehabilitation that can demonstrate long-term motor skill learning may also elucidate important issues related to the modulation of neuronal networks for better resource allocation, the potential for functional cerebral reorganization, and the notion of interventional/adaptive cerebral plasticity, as part of neurorehabilitation (Cifelli and Matthews, 2002).

Chapter 19

Auditory Feedback Control for Improvement of Gait in Patients with Multiple Sclerosis

19.1 Introduction

Positive effects of sensory cues on gait in patients with movement disorders have been reported before, particularly in the context of Parkinson's disease (PD) (Martin, 1967; Azulay *et al.*, 1999). Early attempts to produce such cues artificially have resulted in open-loop systems, producing sensory cues, which are independent of the patient's own motion. Open-loop systems, subject to disturbances, are inherently unstable (Astrom, 1970). Indeed, such systems, displaying visual objects in constant motion (Prothero, 1993), have been found to cause confusion, related to a sense of "falling out of sync" (Baram *et al.*, 2002). Open-loop rhythmic auditory cues have been found to produce different responses to different rhythms and to raise issues of motivational factors (McIntosh *et al.*, 1997; Thaut *et al.*, 1996). In contrast, closed-loop visual feedback signals generated by the patient's own motion have been found to stimulate, stabilize and regulate gait (Baram, 1999). A study on patients with PD (Baram *et al.*, 2002) has shown that while open loop stimuli have adverse effects, particularly dizziness, loss of balance, and even freezing, closed-loop visual feedback cues, responding to the patient's own movement, have a clear positive effect on gait. We recently found a similar effect of closed-loop visual feedback on gait in patients with multiple sclerosis (MS) who suffer from cerebellar ataxia (Baram and Miller, 2006). Previously, auditory feedback has been found to have a positive influence on postural stability in quiet standing tasks

(Dozza *et al.*, 2005). These findings suggest examining the effects of auditory feedback on gait in neurological patients. An apparatus which produces a ticking sound in response to the patient's steps, one tick per step, has been developed. The patient hears the auditory cue produced by steps through an earphone, and can modify the auditory cue by modifying gait. A steady balanced gait will produce a pleasant auditory cue, synchronized with the patient's own steps, rewarding the patient for making the effort.

19.2 Methods

19.2.1 *Subjects*

Fourteen randomly selected outpatients with MS, ten women and four men, diagnosed according to the Poser criteria (Poser *et al.*, 1983) with gait disturbances predominantly due to cerebellar ataxia, participated in the study. Exclusion criteria: auditory dysfunction and gait disturbances due to pronounced muscle weakness, spasticity, sensory ataxia, or significant general fatigue. Disease-related disability was assessed using the Kurtzke Expanded Disability Status Scale (EDSS) (Kurtzke, 1993). In addition, clinical assessment included Cerebellar Functional System Score (CFSS), which grades cerebellar function due to MS on a five-grade score, in which: 0 = normal cerebellar function; 1 = abnormal signs without disability; 2 = mild ataxia; 3 = moderate ataxia; 4 = severe ataxia in all limbs or trunk; 5 = unable to perform coordinated movement due to ataxia.

19.2.2 *Ambulation assessment*

Assessment of patient ambulation, conducted by an impartial independent investigator, was carried out using the ambulation parameters of walking speed (meter/second) and stride length (meters). The control group consisted of 11 healthy individuals. The study was approved by the institutional Helsinki committee, and informed consent was obtained from each of the participants once the nature of the procedures had been fully explained. Clinical and personal data of patients and control groups are presented in Table 19.1.

Table 19.1. Clinical characteristics of patients and control groups.

MS patient group						Control group		
Patient	Sex	Age	DD	EDSS	CFSS	Subject	Sex	Age
1	W	46	18	6	4	1	M	26
2	M	55	24	6	4	2	M	27
3	W	48	3	5	4	3	W	28
4	W	59	5	4	3	4	W	23
5	W	56	1.5	4	2.5	5	W	26
6	M	31	10	4.5	3	6	M	25
7	W	50	6	6	4	7	W	23
8	W	45	5	3.5	3	8	W	23
9	W	49	2	4.5	3	9	M	24
10	W	54	10	4.5	3.5	10	M	28
11	W	54	12	5.5	4	11	W	27
12	W	33	5	5.5	3.5			
13	M	49	21	3.5	2.5			
14	M	51	3	4	3			

19.2.3 *Auditory feedback apparatus and mode of assessment*

Auditory cues were generated by a device that sounds, through a headset or an earpiece, a tick each time the user takes a step. A belt-mounted box, the size and weight of a small cell phone, contains a motion sensor, a micro-controller, and software, which implements an adaptive filter for transforming the user's movement into sound. The user, in turn, adjusts gait so as to produce a balanced rhythmic auditory cue. The expected result is an improved gait pattern.

19.2.4 *Procedure*

All tests were performed at the Multiple Sclerosis Center, Carmel Medical Center, Haifa, Israel, at about the same daytime. Examination of each patient comprised four stages, each consisting of the patient walking a straight track of 10 meters: baseline, device off, device on, and residual effects, as hereinafter:

Stage 1:

The ambulation parameters, baseline walking speed and stride length were measured first, without the device. The patient was verbally instructed to "walk normally". The time to complete the 10-meter track and the number of steps were recorded four times and averaged.

Stage 2:

The device was turned on. The patient was instructed to walk so as to make the auditory cue as rhythmic as possible along the 10-meter track. Walking speed and stride length were measured four times and averaged.

Stage 3:

The device was taken off the patient, who was given a ten-minute break. After the break, the patient was instructed to "walk normally" the 10-meter track without the device. Walking speed and stride length were recorded four times and averaged. The purpose of this stage was to measure the residual short-term therapeutic effect of auditory feedback.

19.3 Results

19.3.1 *Clinical features, baseline parameters, and on-line improvement*

The test results for the patients and the control subjects are summarized in Table 19.2. Baseline walking speed and stride length of patients before using the device are presented. Walking speed and stride length of patients using the device are specified in the "on-line" column, along with percentage changes from "baseline" to "device on". Walking speed and stride length after the break, without the device, are given in the "residual effect" columns, along with percentage changes with respect to the baseline walking speed and stride length

For the patient group, the on-line walking speed improved, on average, 12.84% (standard deviation 18.74%). On-line average improvement in stride length was 8.30% (standard deviation 11.87%).

Table 19.2. Test results for MS patients and for control subjects: walking speed (meter/second) and stride length (meters).

			On-line				Residual			
	Baseline ambulation		Device on		Percentage change		Device off		Percentage change	
Patient no.	Walking speed	Stride length	Walking speed	Stride length	Walking speed	Stride length	Walking speed	Stride length	Walking speed	Stride length
MS patients										
1	0.252	0.256	0.400	0.333	58.73%	30.01%	0.361	0.312	43.25%	21.86%
2	0.647	0.444	0.493	0.435	−19.97%	−15.20%	0.684	0.526	5.72%	18.47%
3	1.043	0.513	0.984	0.541	−5.66%	5.46%	1.211	0.588	16.11%	6.29%
4	1.381	0.690	1.581	0.714	14.48%	3.48%	1.616	0.714	17.02%	3.28%
5	0.880	0.556	1.103	0.597	25.35%	7.37%	1.044	0.588	18.64%	5.76%
6	0.941	0.571	0.982	0.625	4.36%	9.46%	0.933	0.555	−0.85%	−2.80%
7	0.712	0.476	0.786	0.488	10.39%	2.52%	0.798	0.476	12.10%	0.00%
8	1.185	0.588	1.306	0.625	10.21%	6.29%	1.307	0.625	10.30%	6.29%
9	0.871	0.487	0.877	0.500	0.68%	2.67%	0.912	0.500	4.71%	2.67%
10	0.966	0.500	1.067	0.555	10.46%	11.00%	1.104	0.555	14.29%	11.00%
11	0.987	0.526	1.053	0.526	6.69%	0.00%	1.267	0.571	28.37%	8.56%
12	0.616	0.333	0.886	0.444	43.83%	33.33%	0.859	0.426	39.45%	27.93%
13	1.243	0.769	1.362	0.833	9.57%	8.32%	1.815	0.909	46.02%	18.21%
14	1.253	0.690	1.386	0.769	10.61%	11.45%	1.346	0.769	7.42%	11.45%
Control subjects										
1	1.361	0.769	1.298	0.769	−2.27%	0.00%	1.398	0.800	2.66%	4.00%
2	1.340	0.714	1.278	0.714	−6.58%	0.00%	1.407	0.714	5.07%	0.00%
3	1.852	0.833	1.953	0.833	2.54%	0.00%	1.890	0.833	2.08%	0.00%
4	1.576	0.714	1.500	0.769	−6.98%	7.69%	1.637	0.714	3.85%	0.00%
5	1.457	0.800	1.302	0.769	−12.83%	−3.85%	1.517	0.800	4.17%	0.00%
6	1.534	0.833	1.489	0.769	−1.04%	0.00%	1.572	0.769	2.52%	−7.69%
7	1.638	0.769	1.569	0.769	1.33%	0.00%	1.625	0.769	−0.81%	0.00%
8	1.408	0.667	1.420	0.690	−1.35%	3.45%	1.517	0.714	7.74%	7.14%
9	1.202	0.714	1.222	0.714	3.05%	0.00%	1.271	0.714	5.72%	0.00%
10	1.393	0.714	1.173	0.667	−5.92%	0.00%	1.284	0.667	−7.83%	−6.67%
11	1.308	0.667	1.311	0.667	−3.21%	−3.33%	1.329	0.667	1.59%	0.00%

The results for the controls did not show any particular trend in relation to the baseline walking speed or stride length. Furthermore, using the device did not improve performance, due to the burdening effect of wearing the device. Indeed, it was found that the on-line walking speed changed, on average, by −3.02% (standard deviation 4.76%). As in the case of the walking speed, there was no particular trend in the on-line percentage improvement in the stride length as a function of the baseline in the controls. The on-line average change in the stride length was also negligible (0.36%, with standard deviation 3.09%). It can be seen that the level of variation (or standard deviation) in the performance results among patients and among controls is similar to the level of average improvement in performance.

19.3.2 *Percentage improvement*

The percentage improvement in the on-line walking speed of the patients and of the controls with respect to the baseline walking speed is depicted for comparison in Fig. 19.1. Similarly, the residual improvement of the two groups is depicted in Fig. 19.2.

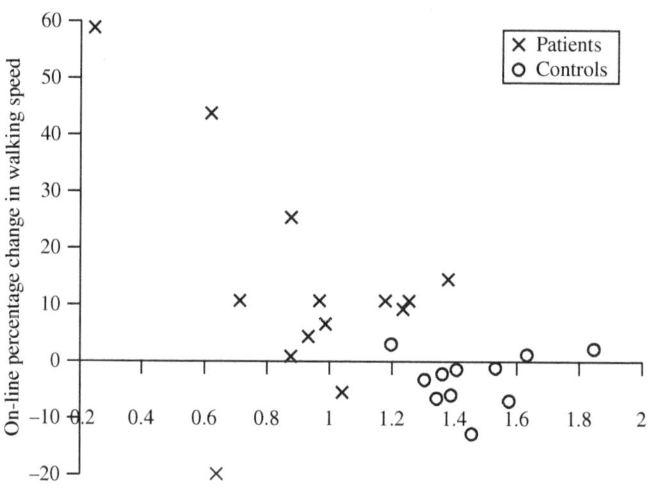

Figure 19.1. On-line percentage improvement in the walking speed of MS patients and control subjects as a function of the baseline walking speed.

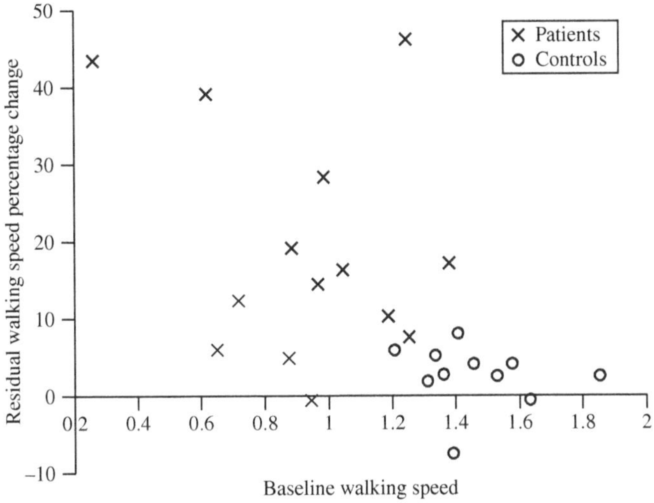

Figure 19.2. Residual (short-term therapeutic) percentage improvement in the walking speed of MS patients and control subjects as a function of the baseline walking speed.

19.3.3 *Residual improvement*

The residual improvement in the walking speed (Fig. 19.2) was, on average, 18.75% (standard deviation 18.53%). Average residual improvement in stride length was 9.93% (standard deviation 9.46%). In contrast, there was no significant residual improvement in the walking speed and the stride length of the controls (2.43% and −0.29%, with standard deviations 4.09% and 4.11%, respectively). Again, it can be seen that the level of variation is close to that of the average improvement.

19.4 Discussion

The mechanism by which auditory signals assist in coordinating rhythmic or sequential movements remains to be determined. There appears to be some evidence that rhythmic sound patterns can increase the excitability of spinal motor neurons via the reticulospinal pathway, thereby reducing the amount of time required for the muscles to respond to a given motor command (DelOlmo and

Cuderio, 2003). There is also increasing evidence that the basal ganglia has an important role in the proper sequencing of complex movements (Palsev *et al.*, 1967). Studies in monkeys have shown that movement-related phasic discharge of pallidal neurons may serve as an internal cue to the supplementary motor area, signaling the end of one movement and allowing the onset of the next (Brotchie *et al.*, 1991). These mechanisms do not necessarily hold the key to movement disorders in MS, where the problem is predominantly in the white matter and not in the gray matter. Furthermore, while the basal ganglia and the pallidum may play a role in certain movement disorders such as Parkinson's disease (Côté and Crutcher, 1991), the connection of the basal ganglia to ataxia, which is predominantly related to brainstem–cerebellar connections, is not clear. Yet, studies of neuronal substrates and mechanisms for auditory motor control may lead to strategies for gait management and rehabilitation in MS.

Open-loop verbal instructional cues (Behrman *et al.*, 1998) and rhythmic auditory cues (Thaut *et al.*, 1996; McIntosh *et al.*, 1997) have been used for gait management in patients with PD. However, open-loop rhythmic cues raise the question of which rhythm would suit a given patient at a given time (for instance a rhythm 10% faster than the baseline (Baram *et al.*, 2002) may suit most but not all patients). In addition, repeated use of rhythmic sound of relatively low complexity is assumed to induce a great amount of redundancy into the perceptual process and thus strongly reduce effective arousal effects related to motivation (Thaut *et al.*, 1996; McIntosh *et al.*, 1997). A study on the role of auditory feedback in maintaining postural balance in stance (Maurer *et al.*, 2003) suggests that auditory feedback increases postural stability in quiet standing tasks and results in a more prominent role for feedback (closed-loop) control over feed-forward open-loop) control. Furthermore, it indicates that the solution proposed by the brain with auditory feedback seems to involve more feedback control for a more stable sway. A comparison with visual feedback control of an upright, quiet stance (Rougier, 1999) suggests non-redundant roles in multi-sensory integration for the control of posture. Whereas vision provides information about the external environment and allows prediction of forthcoming events, auditory information processing time is markedly faster than visual reaction times, making it more important for postural reaction to disturbing stimuli (Rougier, 1999), that is, stability feedback control. It should also be noted that an augmented auditory feedback device, employing a pressure sensitive foot-switch, has been proposed for the treatment of equines (toe-walking) gait in children (Conrad and Bleck, 1980). Auditory feedback influences gait by creating an external cue, which serves as a reference for the patient. However, the feedback nature of this external cue makes it non-imposing, as it is fully controlled by the patient. It creates, on one hand, a constant awareness of gait quality and, on the other, an instantaneous sensory response to changes in gait quality. The patient, by own control, then makes an effort to improve gait quality and is, in turn, informed of any improvement (or deterioration) in gait quality by changes in the auditory cue.

The present study indicates that auditory feedback can help patients with MS control their gait. Patients with baseline performance below the median show a considerably higher improvement than patients above the median. In other words, the more room there is for improvement, the more improvement there is. Compared to our previous study (Baram and Miller, 2006), the present results show some notable differences between the effects of visual and auditory feedback. On average, gait improvement due to auditory feedback was higher than that achieved by visual feedback, which may require more elaborate information processing by the brain (Rougier, 1999). While visual feedback produced a higher improvement in stride length than in walking speed, the improvement in walking speed due to auditory feedback was higher than the improvement in stride length. The higher improvement in stride length caused by visual feedback can be attributed to the bio-feedback effect of visually reaching a given target, that is, the edge of a tile, by extending the foot farther. On the other hand, the higher walking speed caused by auditory feedback reinforces earlier findings that auditory signals reduce reaction time in a voluntary motor task (Conard and Bleck, 1980). Relative to the baseline performance, the residual improvement is even greater than the on-line improvement. This is explained by the fact that there was no off-line training involved. Removing the burdening effect of the device, combined with the learning gained while using the device, maximizes the immediate residual performance. The relatively large short-term residual improvement in patients using auditory feedback is particularly encouraging, as it suggests a therapeutic potential. The improvements in performance seem particularly noteworthy when compared to the results obtained for the controls, which showed no meaningful improvement. This suggests that an improved device, reducing the burdening effect, and extended training, may further improve patient performance.

Audio-based assistive technology may have special importance in reinforcement of functions in patients who suffer from visual impairment (and are therefore not candidates for the visual system) and patients suffering from proprioceptive input problems. Since the auditory and the vision feedback channels seem to produce somewhat complementary functional enhancement properties, and since, as suggested by earlier studies (Rougier, 1999), multi-sensory integration may be key to motor stabilization and control, future studies may examine the effects of simultaneous visual and audio feedback through various artificial means on walking abilities of MS and other neurological patients. Future studies may also examine the combined long-term therapeutic effects of visual and audio feedback on gait in movement-disorder patients. The analysis of larger groups of patients would facilitate a comparison between patient subcategories, responders, and non-responders to sensory feedback, in particular, and a determination of specific predictive factors.

Chapter 20

The Gait-Entrainment Edge of Glide-Symmetric Virtual Visual Feedback

20.1 Introduction

As noted in other chapters of this book, visual feedback has been found to affect the walking abilities of patients with neurological disorders. While early studies employed equally spaced transverse lines drawn on the ground (Martin, 1967), later reports noted similar effects by other visual patterns, such as a checkerboard-like tile arrangement (Sacks, 1973). In a previous study (Baram and Miller, 2006), we found that the improvement in the walking ability of multiple sclerosis (MS) patients training with a virtual-reality device that displayed earth-stationary checkerboard-like tiles lasted beyond the use of the device. While this finding suggests that training with such cues may be an effective treatment modality for neurological gait disorders, the significance of the particular visual pattern used has remained an open question. Specifically, a question of interest is whether there is a fundamental difference between transverse lines and checkerboard-like tile arrangement that might affect the level of gait improvement. An examination of the human locomotion mechanism reveals that it transforms the reflection-symmetry of the limbs into a somewhat more compound symmetry, called "glide-reflection" (Livio, 2005).

As illustrated by Fig. 20.1(b), the glide reflection geometry consists of both reflection and shifting. Unlike transverse lines (Fig. 20.1(a)) that have no distinct symmetry (although any line segment is reflection-symmetric about its middle point, there is no

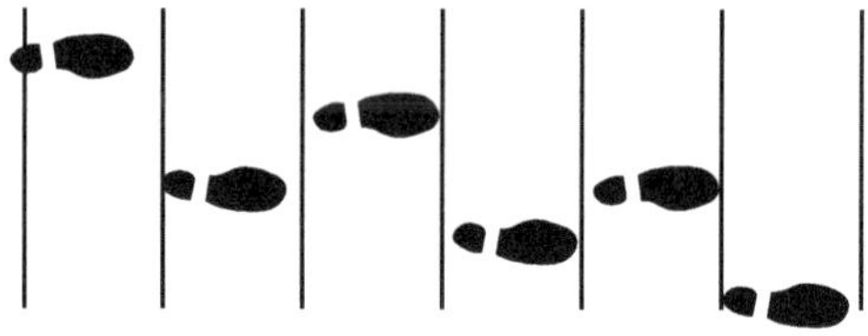

(a)

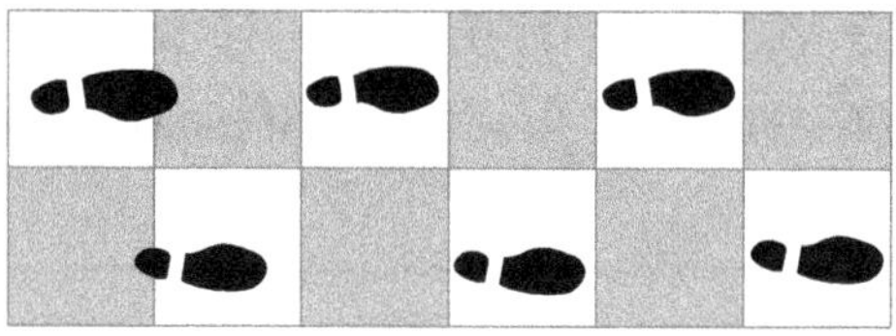

(b)

Figure 20.1. Walking with (a) transverse lines and (b) checkerboard tiles feedback. Following a few initial steps, the steps synchronize with the lines or the tiles. The transition from one-dimensional (a) to two-dimensional (b) novelty and vigilance (see Chapter 14) results in lateral drift elimination.

visible reflection surface), the geometry of the checkerboard tile arrangement (Fig. 20.1(b)) has a distinct glide-reflection symmetry. The checkerboard tile arrangement is matched, then, to the geometry of human locomotion.

In the context of walking, this geometry represents not only, as transverse lines, the reach of the foot in every step, but also the transition from one foot to the other. We hypothesize that memorization of the visual feedback cue is a part of the gait training process, which would imply that training with the checkerboard tile arrangement will result in a higher level of improvement than training with transverse lines. The following study (Baram and Miller, 2010) was set to test this hypothesis.

20.2 Methods

20.2.1 *Subjects*

Twenty-one randomly selected MS outpatients, with gait disturbances predominantly due to cerebellar ataxia, participated in the study. Exclusion criteria: considerable visual deficit not compensated by correction, ocular movement dysfunction (e.g., diplopia, nystagmus), and gait disturbances due to pronounced muscle weakness, spasticity, sensory ataxia, or considerable general fatigue. Disease-related disability was assessed using the Kurtzke Expanded Disability Status Scale (EDSS) (Kurtzke, 1993). The patients were randomly divided into two disjoint groups, one including 10 patients, the other 11. The first group was to be trained with visual cues consisting of transverse lines, the other with visual cues of a checkerboard-like tile arrangement.

20.2.2 *Ambulation assessment*

Assessment of patient ambulation, conducted by an impartial independent investigator, was carried out using walking speed (meter/second) and stride length (meters), calculated directly from measurements of time and number of steps needed to complete a fixed track. The study was approved by the institutional Helsinki committee and informed consent was obtained from each of the participants once the nature of the procedures had been fully explained.

20.2.3 *Virtual reality apparatus*

Visual cues were generated by a Virtual Reality (VR) apparatus (Baram *et al.*, 2002; Baram, 2004). The display, attached to the eyeglasses frame, provided patients of the first group with a display of transverse lines and patients of the second group with a virtual tiled floor in a checkerboard arrangement, both responding dynamically to the patient's own motion, much like a real floor, fixed in space. The side-length of the tiles in the second arrangement was equal to the spacing between the transverse lines in the first.

20.2.4 *Procedure*

All tests were performed at the Multiple Sclerosis Center, Carmel Medical Center, Haifa, Israel, at about the same time in the morning.

Examination of each patient comprised three stages:

Stage 1.

Baseline walking speed and stride length without the device were measured. The patient was verbally instructed to walk along a straight track of 10 meters. The time to complete the track and the number of steps were recorded four times and averaged.

Stage 2.

The device was placed on the patient, and turned on. The patient was instructed to walk four times along the 10-meter track for the purpose of training. The patient was asked to imagine, while walking, stepping between the virtual lines or on the virtual checkerboard-like tiles, as applicable.

Stage 3.

The device was taken off the patient, who was given a ten-minute break. After the break, the patient was instructed to walk the 10-meter track without the device. Walking speed and stride length were recorded four times and averaged. The purpose of this stage was to measure the residual short-term therapeutic effect of training with each of the two specified visual feedback cues.

20.3 Results

The test results for Groups 1 (training with transverse lines) and 2 (training with checkerboard tiles) are summarized in Tables 20.1 and 20.2, respectively, where clinical and personal data of patients are presented first. Baseline walking speed and stride length of patients before using the device are given in the "Baseline" columns. Walking speed and stride length after training with the device are given in the "After training" columns, and the percentage changes with respect to the baseline walking speed and stride length are given in the "Percentage change" columns. The average improvement in Group 1 was $7.79 \pm 4.24\%$ in walking speed and $7.20 \pm 3.92\%$ in stride length. The average improvement in Group 2 was $21.09 \pm 18.39\%$ in walking speed and $12.99 \pm 11.72\%$ in stride length.

Table 20.1. Test results for patients of Group 1 (training with transverse lines; DD = disease duration in years; EDSS = Expanded Disability Status Scale; walking speed in meter/second, stride length in meters).

Patient	Sex	Age	DD	EDSS	Baseline		After training		Percentage change	
					Walking speed	Stride length	Walking speed	Stride length	Walking speed	Stride length
1	W	59	7	3	0.755	0.426	0.808	0.488	7.02%	14.55%
2	W	33	4	4	0.820	0.455	0.870	0.488	6.10%	7.25%
3	M	44	19	3.5	1.196	0.714	1.422	0.769	18.90%	7.70%
4	W	39	14	5.5	1.140	0.606	1.252	0.625	9.82%	3.14%
5	M	16	1	3	1.186	0.690	1.259	0.690	6.16%	0.00%
6	W	51	6	2.5	1.191	0.645	1.277	0.667	3.02%	3.40%
7	W	21	1	3.5	1.211	0.588	1.257	0.645	3.80%	9.69%
8	W	45	10	5	0.497	0.417	0.525	0.426	5.63%	8.67%
9	M	39	6	3	0.707	0.465	0.772	0.500	19.19%	7.53%
10	W	43	12	5	1.168	0.606	1.264	0.667	8.22%	10.07%

Table 20.2. Test results for patients of Group 2 (training with checkerboard-like tiles; DD = disease duration in years; EDSS = Expanded Disability Status Scale; walking speed in meter/second, stride length in meters).

Patient	Sex	Age	DD	EDSS	Baseline		After training		Percentage change	
					Walking speed	Stride length	Walking speed	Stride length	Walking speed	Stride length
1	W	24	5	6	1.093	0.555	1.153	0.588	5.49%	5.95%
2	W	20	5	4.5	1.316	0.645	1.440	0.667	9.42%	3.41%
3	M	54	23	6	0.577	0.425	0.708	0.513	22.70%	20.71%
4	M	30	2	5.5	1.072	0.625	1.204	0.667	12.31%	6.72%
5	W	34	3	4.5	0.808	0.476	0.962	0.513	16.32%	7.77%
6	M	50	1	3.5	1.194	0.667	1.508	0.769	26.30%	15.29%
7	M	67	4	6.5	0.178	0.162	0.180	0.172	1.12%	6.17%
8	W	50	25	2	1.272	0.625	1.278	0.606	−0.93%	−3.04%
9	W	48	6	5.5	0.480	0.357	0.666	0.417	38.75%	16.81%
10	W	46	18	3.5	0.321	0.290	0.442	0.370	37.69%	42.31%
11	W	47	3	5	0.575	0.385	0.936	0.465	62.78%	20.78%

20.4 Discussion

It has been suggested that visual feedback of transverse lines may improve walking in patients with Parkinson's disease (PD) (Martin, 1967). Similar effects caused by tiled floor patterns have been reported (Sacks, 1973). Consequently, the stabilizing effect of such information has been shown (Baram, 1999; Baram *et al.*, 2002), and the delivery of a tiled floor pattern, responding dynamically to the patient's own motion through a head-mounted VR device, has been suggested (Baram, 2004). The gait quality of PD patients (Baram *et al.*, 2002) and MS patients (Baram and Miller, 2006) aided by a portable virtual reality device (Baram, 2004) improved considerably. Yet, no particular significance has been associated with the geometry of the cue used for visual feedback. The results of the present study show that the choice of glide-symmetric geometry is of fundamental significance, having a substantial impact on the level of gait improvement. Specifically, a checkerboard-like tile pattern was found to have a considerably greater effect on gait improvement than a transverse line pattern. Furthermore, the level of improvement facilitated by the checkerboard-like tile pattern is inversely related to the level of impairment. We propose that the difference between the effects of the two patterns is associated with the types of symmetry that they convey. Unlike transverse lines, which have no distinct symmetry, the checkerboard tile arrangement has a distinct glide-reflection symmetry, matched to the human locomotion mechanism. Besides the physiotherapeutic effect of gait training, the visual feedback generated by locomotion through the VR device appears to be memorized, controlling the orderly transition of the legs in a glide-symmetric fashion. This long-term effect makes training with glide-symmetric visual feedback an effective treatment modality for patients with gait impairment.

Chapter 21

Gait Improvement in Patients with Cerebral Palsy by Visual and Auditory Feedback

21.1 Introduction

Cerebral palsy (CP) is a group of prenatal disorders that present cognitive, motor, and sensory disabilities (Woollacott and Burtner, 1996). Classification is difficult as each of these disorders and disabilities appears within a wide range of independent severity. The most prominent feature is motor function disorder, associated with a general reduction in electromyogram (EMG) activity, selected muscle weakness, and afferent inhibition (Berger *et al.*, 1982; Berger, 1998). The motor impairment syndromes seen in CP are secondary to lesions or anomalies in brain development, inhibiting normal automatic and voluntary responses and allowing for the development of dysfunctional ones (Woollacott and Burtner, 1996; Berger *et al.*, 1982). Bilateral and hemiplegic spasticity give rise to difficulties in posture, balance, and gait control (Berger *et al.*, 1982; Berger, 1998; van der Heide and Hadders-Algra, 2005). Mobility-related handicaps can considerably limit daily activity, affecting individual well-being and social interaction, and, therefore, have a major impact on overall patient development.

The treatment of CP involves physical therapy, occupational therapy, and sports, targeted at specific conditions. Upper extremity of arm reaching tasks was found to improve in CP patients training in a computer-generated environment (Chen *et al.*, 2007). EMG feedback was found to be potentially beneficial for gait improvement in patients with CP-related impairment, specifically dynamic equinus deformity (Dursun *et al.*, 2004) and spastic hemiplegia (Colborne *et al.*, 1994). Training with sensory feedback has been found to improve motor function. Auditory feedback, provided by load-sensitive insole inserted in the hemiparetic side shoe, was found to have a positive effect on gait symmetry in CP (Seeger *et al.*, 1981), but maintenance strategies for this effect are yet to be determined (Seeger and Caudrey, 1983). Balance training with visual feedback was found to improve stance in hemiplegic CP, with a secondary positive effect on gait (Ledebt *et al.*, 2005).

Walking over earth-stationary transverse lines has been found to improve gait in patients with Parkinson's disease (PD) (Martin, 1967; Azulay *et al.*, 1999). An analytic study of the effects of real-world visual cues on the regulation and stabilization of gait (Baram, 1999) has led to the development of a wearable virtual reality (VR) apparatus (Batam, 2004), which incorporates the patient's body motions in the display, so that the visual cues appear to be earth-stationary. Training with this apparatus was found to improve gait in patients with PD (Baram *et al.*, 2002) and patients with multiple sclerosis (MS) (Baram and Miller, 2006). Similarly, an auditory feedback device, sounding a clicking sound in response to each step, was found to improve gait in patients with MS (Baram and Miller, 2007). In the present study (Baram and Lenger, 2012), we employed these devices to study the effects of training with sensory feedback cues on gait in patients with CP.

21.2 Materials and methods

Study population

Ten patients complaining about gait disturbances due to CP trained with visual feedback, while ten such patients trained with auditory feedback. While six of the patients participated in both groups, the auditory feedback training and tests were performed one week after the visual feedback training and tests, so as to reduce cross effects of the brief 20-minute training periods. Seven age-matched healthy controls trained with visual feedback, while eight such controls trained with auditory feedback. Seven of the controls participated in both groups, but the visual and the auditory training and tests were separated by 20 days. The study was approved by the Helsinki Committee of the Ministry of Health, Israel, and informed consent was obtained from a parent of each of the participants once the nature of the procedures had been fully explained.

Materials

Visual cues were generated by a VR apparatus shown on the left-hand side of Fig. 21.1. The display, attached to the eyeglasses frame, provided the patients with an image of transverse lines, responding dynamically to the patient's own motion, much like a real floor fixed in space. Auditory feedback cues were generated by a device shown on the right hand side of Fig. 21.1, which delivers through earphones a clicking sound in response to each step. Both devices employ an inertial motion sensor and a computing unit enclosed in a belt-mounted box. Assessment of patient ambulation was based on walking speed and stride length, which are calculated directly from measurements of time and number of steps needed to complete a fixed track.

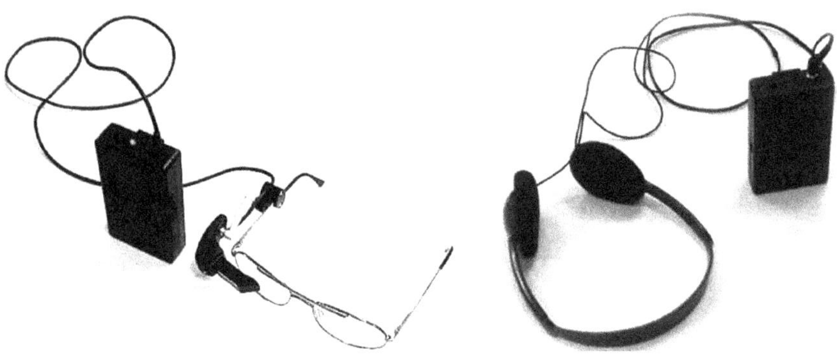

Figure 21.1. (left) Visual feedback device used in tests. (right) Auditory feedback device used in CP tests.

21.2.1 *Procedure*

All tests were performed at the Organization for Cerebral Palsy Victims in Israel (ILAN) Sports Center for Handicapped Children, Kiryat Haim, Israel, at about the same time in the afternoon. Examination of each patient comprised three stages:

Stage 1

Baseline walking speed and stride length without the device were measured. The patient was verbally instructed to walk along a straight track of 10 meters. The time to complete the track and the number of steps were recorded four times and averaged.

Stage 2

The device was placed on the patient and turned on. The patient was instructed to walk four times along the 10-meter track for the purpose of training. Patients training with visual feedback were asked to imagine, while walking, stepping between the virtual transverse lines. Patients training with auditory feedback were asked to maintain, by controlling their gait pattern, a rhythmic auditory cue.

Stage 3

The device was taken off the patient, who was given a 20-minute break. After the break, the patient was instructed to walk the 10-meter track without the device. Walking speed and stride length were recorded four times and averaged. The purpose of this stage was to measure the residual short-term therapeutic effect of training with the sensory feedback cue.

21.3 Results

21.3.1 *Visual feedback*

The test results for patients training with visual feedback are summarized in Table 21.1. The standard deviation for the four trials averaged was smaller than 10% of the mean value reported for each of the patients. The average improvement in walking speed was 20.70% ± 36.06% (the second number representing standard

Table 21.1. Test results for CP patients training with visual feedback (age in years, walking speed in meter per second, and stride length in meters).

Patient			Baseline		After training		Percentage change	
Name	Sex	Age	Walking speed	Stride length	Walking speed	Stride length	Walking speed	Stride length
1	F	10	1.285	0.555	1.265	0.571	−1.56%	2.88%
2	M	18	1.419	0.741	1.508	0.769	6.27%	3.78%
3	F	11	1.439	0.606	1.357	0.625	−5.70%	3.14%
4	F	7	0.44	0.328	0.648	0.385	47.27%	17.38%
5	F	26	0.945	0.625	1.052	0.667	10.15%	6.72%
6	F	9	0.401	0.27	0.369	0.263	−7.98%	−2.59%
7	M	6	0.81	0.377	0.939	0.476	15.93%	26.26%
8	F	12	0.687	0.476	1.466	0.571	113.40%	19.96%
9	M	15	1.224	0.588	1.513	0.645	23.61%	9.69%
10	F	19	1.362	0.69	1.575	0.69	15.64%	0.00%

deviation). This result was found to be statistically significant in the sense that the hypothesis that the average improvement found corresponds to the given sample was accepted with a confidence level of 0.95. For participants with baseline walking speed below the median (1.08 meter/second), improvement was considerably higher (35.75% ± 47.76%) than for participants with baseline walking speed above the median (7.65% ± 12.85%). The hypothesis was further affirmed by the fact that, while the hypotheses that the average improvements of both groups correspond to the respective samples were accepted at the 0.95 confidence level, the hypothesis that the average improvement of patients with below-median baseline performance corresponds to the above-median sample was rejected at the 0.95 confidence level. Average improvement in stride length was 8.72% ± 9.47%, with a 0.95 confidence level. For participants with baseline stride length below the median (0.57 meters), the improvement was 12.78% ± 12.13%, and for participants with a baseline stride length above the median, the improvement was 4.67% ± 3.69%. While both results were confirmed at the 0.95 confidence level, the hypothesis that the average improvement of patients with below-median baseline performance corresponds to the above-median sample was rejected at the 0.95 confidence level. For participants of age below the median (11.5 years), improvement in walking speed was 9.59% ± 23.06%, while for participants of age above median, the improvement was 33.81% ± 44.96%. For participants of age below the median, improvement in stride length was 9.41% ± 11.97%, while for participants of age above the median, the improvement was 8.03% ± 7.57%. However, in contrast to the dependence of the test significance on the baseline performance, no such dependence was found with respect to the education level in either the walking speed or the stride length results.

21.3.2 *Auditory feedback*

The test results for patients training with auditory feedback are summarized in Table 21.2. As in the case of visual feedback, the standard deviation for the four trials averaged was smaller than

Table 21.2. Test results for CP patients training with auditory feedback.

Patient			Baseline		After training		Percentage change	
Name	**Sex**	**Age**	**Walking speed**	**Stride length**	**Walking speed**	**Stride length**	**Walking speed**	**Stride length**
1	F	10	1.224	0.513	1.342	0.555	9.64%	8.19%
2	M	18	1.262	0.645	1.259	0.645	6.27%	0.00%
3	M	6	0.851	0.417	1.266	0.5	48.77%	19.90%
4	F	7	0.702	0.333	0.942	0.385	34.19%	15.62%
5	F	26	0.892	0.555	0.914	0.588	2.47%	5.95%
6	F	11	1.171	0.526	1.406	0.588	20.07%	11.79%
7	M	5	0.841	0.351	1.006	0.4	19.62%	13.96%
8	M	5	1.183	0.435	1.206	0.444	1.94%	2.07%
9	F	12	0.543	0.426	0.629	0.476	15.84%	11.74%
10	F	11	0.739	0.455	1.445	0.667	95.53%	46.59%

10% of the mean value reported for each of the patients. The average improvement in walking speed was 25.43% ± 28.65% with a confidence level of 0.95. For participants with baseline walking speed below the median (0.87 meter/second), the improvement was considerably higher (42.79% ± 32.23%) than for participants with baseline walking speed above the median (8.08% ± 7.39%). While the hypotheses that the average improvements of both groups correspond to the respective samples were accepted at the 0.95 confidence level, the hypothesis that the average improvement of patients with below-median baseline performance corresponds to the above-median sample was rejected at the 0.95 confidence level. Average improvement in stride length was 13.58% ± 13.10% with a confidence level of 0.95. For participants with baseline stride length below the median (0.45 meters), the improvement was 13.66% ± 6.63%, and for participants with a baseline stride length above the median, the improvement was 14.50% ± 8.44%. While the hypotheses that the average improvements of both groups correspond to the respective samples were accepted at the 0.95 confidence level, the hypothesis that the average improvement of patients with below-median baseline performance corresponds to the above-median sample was rejected at the 0.95 confidence level. For participants of age below the median (10.5 years), the improvement in walking speed was 22.83% ± 18.86%, while for participants of age above the median, the improvement was 28.04% ± 38.39%. For participants of age below the median, the improvement in stride length was 11.95% ± 6.94%, while for participants of age above the median, the improvement was 15.21% ± 18.20%. As in the case of the visual feedback, in contrast to the dependence of the test significance on the baseline performance, no such dependence was found with respect to the education level in either the walking speed or the stride length results.

21.3.3 *Controls results*

The results for the age-matched control groups training with visual and auditory feedback are given in Tables 21.3 and 21.4, respectively. For the controls who trained with visual feedback, the

Table 21.3. Test results for controls training with visual feedback.

Patient			**Baseline**		**After training**		**Percentage change**	
Name	**Sex**	**Age**	**Walking speed**	**Stride length**	**Walking speed**	**Stride length**	**Walking speed**	**Stride length**
1	F	13	1.470	0.714	1.533	0.714	4.29%	0%
2	M	13	1.493	0.769	1.592	0.833	6.63%	8.32%
3	M	12	1.482	0.714	1.477	0.714	−0.34%	0%
4	F	12	1.460	0.728	1.487	0.769	1.85%	5.63%
5	F	12	1.304	0.625	0.987	0.468	−24.31%	−25.12%
6	M	13	1.358	0.714	1.295	0.690	−4.64%	−3.36%
7	M	12	1.557	0.769	1.551	0.728	−0.39%	−5.33%

Table 21.4. Test results for controls training with auditory feedback.

Patient			Baseline		After training		Percentage change	
Name	**Sex**	**Age**	**Walking speed**	**Stride length**	**Walking speed**	**Stride length**	**Walking speed**	**Stride length**
1	F	13	1.607	0.773	1.667	0.702	3.73%	−9.18%
2	M	13	1.667	0.785	1.607	0.801	−3.60%	2.04%
3	M	12	1.75	0.817	1.690	0.716	−3.43%	−12.36%
4	F	12	1.833	0.728	1.667	0.714	−9.06%	−1.92%
5	F	12	1.547	0.678	1.778	0.714	14.93%	5.31%
6	M	13	1.428	0.690	1.547	0.728	8.33%	5.51%
7	M	12	1.667	0.741	1.547	0.690	−7.19%	−5.63%
8	F	14	1.667	0.769	1.607	0.769	−3.60%	0%

average improvement was $-2.41\% \pm 9.54\%$ in the walking speed and $-2.84\% \pm 10.11\%$ in the stride length. For the controls who trained with auditory feedback, the average improvement was $0.01\% \pm 7.73\%$ in the walking speed and $-2.03\% \pm 6.15\%$ in the stride length. In addition to the small average changes in performance and the relatively large variances, it can be seen that in both the visual and the auditory channels, about half of the controls showed a small improvement, while the others showed a small deterioration in performance. It might be concluded that, in general, neither the visual nor the auditory feedback had any measurable effect on the walking qualities of the controls.

21.4 Discussion

The results of the present study support the potential role of sensory feedback-based strategies as rehabilitation modalities and substantiate their specific implementation in efforts to alleviate, improve, and restore mobility in patients with gait disturbances due to CP. For patients using visual feedback, baseline walking speed and stride length were found to be good predictors of improvement, with higher improvement associated with lower baseline performance. For patients using such cues, age was found to be a good predictor of improvement in walking speed, but not in stride length, with significantly higher improvement in walking speed found in patients of older age. For patients using auditory feedback, baseline performance was found to be a good predictor of improvement in walking speed, but not in stride length. Patients of older age showed a somewhat higher improvement than younger patients after training with auditory cues. The significance of these results is enhanced by the results obtained for age-matched healthy controls, who showed no improvement in their walking abilities following training with either visual or auditory feedback (as the performance of these individuals can be assumed to leave little room for improvement and the slight deterioration may be attributed to the burdening effect of the devices). Future studies may explore the long-term therapeutic effects of training with sensory feedback in larger cohorts of patients of various clinical subcategories, possibly in comparison to other treatment modalities. An extended-treatment course should be implemented and its effects examined in terms of both the specific functional performance and long-term skill learning. Research that can link sensory feedback-enhanced rehabilitation to long-term learning would have particular importance in CP, characterized by damage of coordinated inter-neuronal networks leading to associated dysfunction. It might be hypothesized that gait improvement in CP patients employing sensory feedback is achieved through the enhancement of neuronal connections bypassing deficient pathways. However, the examination of this and other possible mechanisms for gait improvement and rehabilitation is left for further research. Sensory feedback-aided motor skill learning also may reveal interventional mechanisms for better resource allocation, functional reorganization, and the role of cerebral plasticity in neurorehabilitation.

Chapter 22

A Brief Study of Virtual Reality Feedback for Gait Improvement in Patients with Idiopathic Senile Gait Disorders and Patients after Stroke

It has been demonstrated that sensory stimuli may improve the walking abilities of patients with movement disorders. In particular, earth-stationary visual cues were found to have a beneficial effect on gait in patients with Parkinson's disease (PD) (Martin, 1967; Azulay *et al.*, 1999). The closed-loop effects of real-world visual cues on the regulation and stabilization of gait were demonstrated in a previous study (Baram, 1999). This has led to the development of a wearable closed-loop virtual reality (VR) apparatus (Baram, 2004), which employs an inertial feedback system to incorporate the patient's body motions in the display, creating a visual sensation of moving with respect to earth-stationary tiles. The visual feedback device was found to be more effective for gait improvement in patients with PD than the open-loop version, displaying fixed-velocity motion (Baram *et al.*, 2002). The device was also found to improve gait in patients with multiple sclerosis (MS) (Baram and Miller, 2006). In the present study, this device was employed to study the effects of visual feedback cues on gait in patients with idiopathic senile gait (SG), some of whom had previous history of strokes (PS) (Baram *et al.*, 2010).

The Israel Ministry of Health and the institutional Helsinki committees approved the study, and informed consent was obtained from each participant once the nature of the procedure was explained. Twenty-one randomly selected old-age home residents suffering from SG, including six with PS, participated in the study. Exclusion criteria were considerable visual deficit not compensated-for by correction; ocular movement dysfunction (e.g., diplopia, nystagmus); and gait disturbances due to pronounced muscle weakness, lower limb spasticity, sensory ataxia, or considerable general fatigue. Visual cues were generated by the closed-loop VR device shown in Fig. 22.1. Tests were performed at two old-age homes in Haifa, Israel, at approximately the same time in the morning. Examination of each patient comprised two stages, each consisting of the patient walking a straight 10-meter track four times. In Stage 1, the patient was verbally instructed to "walk normally" without the device. The time to complete the 10-meter track and the number of steps were recorded. In Stage 2, the device was put on the patient and turned on. The patient was instructed to walk on the virtual tiles while imagining walking on real tiles and attempting to reach the next tile in each step. The time to complete the 10-meter track and the number of steps were recorded. Walking speed (m/s) and stride

Figure 22.1. Visual-feedback virtual reality device used in tests.

length (meters) were calculated from the time and the number of steps needed to complete the 10-meter track. The results were averaged for each stage.

Thirteen patients improved their walking speed, stride length, or both; nine improved their walking speed, stride length, or both by more than 10%. In patients with baseline performance above the median, the average improvement was considerably higher ($6.3 \pm 12.6\%$ in walking speed and $6.4 \pm 11.1\%$ in stride length) than in patients with baseline performance below the median ($-0.7 \pm 22.7\%$ in walking speed and $3.4 \pm 11.3\%$ in stride length). Four of six patients with PS improved their walking speed, stride length, or both by more than 10%. Although all four patients with left hemisphere vascular accident improved their gait, neither patient with right hemisphere vascular accident improved. In the patients with PS with baseline performance above the median, the improvement was considerably higher ($13.2 \pm 6.0\%$ in walking speed and $16.6 \pm 4.7\%$ in stride length) than in patients with baseline performance below the median ($-9.9 \pm 27.0\%$ in walking speed and $-7.7 \pm 12.3\%$ in stride length). The VR visual feedback cues did not improve gait in patients with vascular risk factors but without a history of PS. Education was a relatively good predictor of improvement (e.g., for the 3 patients with 8 years of schooling, the average improvement was $-8.8 \pm 23.8\%$ in walking speed and $-4.7 \pm 15.3\%$ in stride length; for the 4 patients with 12 years of schooling, the average improvement was $-1.8 \pm 26.8\%$ in walking speed and $3.8 \pm 9.7\%$ in stride length; and for the 2 patients with 20 years of schooling, the average improvement was $6.7 \pm 0.5\%$ in walking speed and $14.5 \pm 12.6\%$ in stride length).

The results of the present study indicate that it is possible to use VR cues, superimposed on the real world in a closed-loop fashion, to help patients with senile gait and, in particular, patients with a history of strokes, control their gait. The degree of improvement was found to be related to baseline walking speed and stride length and to the level of education. Future studies should explore the long-term effects of training with such cues for therapeutic and rehabilitation purposes.

Epilogue

The concept of motion control, traditionally associated with the design, construction, and operation of moving mechanical objects, has, over the years, become a rather abstract mathematical discipline. Termed control theory, or systems theory, the corresponding mathematical intricacies have become largely inaccessible to non-mathematicians. Yet, as shown in this book, resorting to certain well established mathematical foundations, namely, probability, statistics, and information, can shed light on highly efficient motion control implementations. A case in point is the model proximity measure (Baram, 1976), which facilitates the optimal selection of a control system model out of a given multi-modal set of control systems for best performance. While computationally more demanding than a single-model design (see, e.g., Chapter 11), this approach appears to be highly useful in certain applications, such as wide-envelope aircraft control system design (see, e.g., Chapter 12). Yet, there appears to be a host of other applications that would highly benefit from such designs. One such application, not immediately conceived in either biological or technological terms, is that of gait entrainment in the neurologically impaired. The high variability of motion impairments in neurological patients can be resolved by on-line model selection, individually adapting to patient's needs. Such model selection, facilitating on-going sensation of virtually generated vision and sound, represents a pseudo-natural, highly valuable cortical ability. The gift of technologically generated virtual sensation, experimentally and clinically tested, not only represents a new direction for motion under neurological impairment, but also emphasizes the many benefits of the relevant mathematics.

Appendix of Theorem Proofs

Proofs of original theorems, corollaries, and lemmas presented in the book are gathered here so as to allow for an easier flow of reading within book chapters.

Proof of Theorem 2.7:

$$I_n(*;s) = -E_*^{U_{n-1}} \log \mathrm{h}_*^s \left(z_n \left| Z^{n-1}\right.\right)$$

Using the inequality

$$\log a \leq a - 1; \ \log a \text{ if and only if } a = 1 \tag{A.1}$$

we get

$$I_n(*;s) \geq 1 - E_*^{U_{n-1}} \mathrm{h}_*^s \left(z_n \left| Z^{n-1}\right.\right) = 0 \text{ a.e.} \tag{A.2}$$

where the second equality follows from Eq. (2.21). To show that equality holds only if $\mathrm{f}_s\left(z_n \left| Z^{n-1}\right.\right) = \mathrm{f}_*\left(z_n \left| Z^{n-1}\right.\right)$ a.e. (sufficiency is trivial).

Suppose that

$$I_n(*;s) = I - E_*^{U_{n-1}} \mathrm{h}_*^s \left(z_n \left| Z^{n-1}\right.\right) \text{ a.e.}$$

i.e.,

$$E_*^{U_{n-1}} \left\{\mathrm{h}_*^s \left(z_n \left| Z^{n-1}\right.\right) - \log \mathrm{h}_*^s \left(z_n \left| Z^{n-1}\right.\right) - 1\right\} = 0 \text{ a.e.}$$

By Eq. (2.0) we then have

$$\int \left[\mathrm{h}_*^s\left(z_n\left|Z^{n-1}\right.\right)-\log \mathrm{h}_*^s\left(z_n\left|Z^{n-1}\right.\right)-2\right]\mathrm{dP}_*=0 \qquad \text{(A.3)}$$

Equations (A.3) and (A.1) together give

$$\mathrm{h}_*^s\left(z_n\left|Z^{n-1}\right.\right)=1 \text{ a.e.}$$

or

$$\mathrm{f}_s\left(z_n\left|Z^{n-1}\right.\right)=\mathrm{f}_*\left(z_n\left|Z^{n-1}\right.\right) \text{ a.e.} \qquad \text{(A.4)}$$

Hence, equality in Eq. (A.2) holds if and only if Eq. (A.4) holds. Similarly, since $I_n(*;s)=0$ a.s., which, as shown above, occurs if and only if

$$\mathrm{f}_s\left(z_n\left|Z^{n-1}\right.\right)=\mathrm{f}_*\left(z_n\left|Z^{n-1}\right.\right) \text{ a.e.} \qquad \square$$

Proof of Theorem 2.8: To prove that $\left|\bar{I}_n(s,t)\right|$ is a pseudo-metric on S for each n, we have to show (see Section 2.6) that for each n, it satisfies the following conditions

(*i*) $\left|\bar{I}_n(s;s)\right|=0$ for any $s\in S$

(*ii*) $\left|\bar{I}_n(s;t)\right|=\left|\bar{I}_n(t;s)\right|$ for any $s,\ t\in S$

(*iii*) $\left|\bar{I}_n(s;t)\right|\le\left|\bar{I}_n(s;r)\right|+\left|\bar{I}_n(r;t)\right|$ for any $s,t,r\in S$

We have

$$\mathrm{h}_s^s\left(z_n\left|Z^{n-1}\right.\right)=1$$

Hence

$$\log \mathrm{h}_s^s\left(z_n\left|Z^{n-1}\right.\right)=0$$

Then also

$$E\log \mathrm{h}_s^s\left(z_n\left|Z^{n-1}\right.\right)=0$$

and (i) follows. Also

$$\bar{I}(s,t) = -\bar{I}(t,s)$$

and (ii) follows.

Condition (iii) is proved as follows

$$\begin{aligned}
&\left|\bar{I}_n(s;r)\right| + \left|\bar{I}_n(r;t)\right| \\
&\quad = \left|E_* \log \mathrm{h}_r^s\left(z_n \left|Z^{n-1}\right.\right)\right| + \left|E_* \log \mathrm{h}_t^r\left(z_n \left|Z^{n-1}\right.\right)\right| \\
&\quad = \left|E_* \log \mathrm{f}_s\left(z_n \left|Z^{n-1}\right.\right)\right| - \left|E_* \log \mathrm{f}_r\left(z_n \left|Z^{n-1}\right.\right)\right| \\
&\qquad + \left|E_* \log \mathrm{f}_r\left(z_n \left|Z^{n-1}\right.\right)\right| - \left|E_* \log \mathrm{f}_t\left(z_n \left|Z^{n-1}\right.\right)\right| \\
&\quad \geq \left|E_* \log \mathrm{f}_s\left(z_n \left|Z^{n-1}\right.\right)\right| - \left|E_* \log \mathrm{f}_t\left(z_n \left|Z^{n-1}\right.\right)\right| \\
&\quad = \left|E_* \log \mathrm{h}_t^s\left(z_n \left|Z^{n-1}\right.\right)\right| = \left|\bar{I}_n(s;t)\right|
\end{aligned}$$

If, in addition to (i), (ii), and (iii), $\left|\bar{I}_n(s;t)\right|$ satisfies

(iv) $\left|\bar{I}_n(s;t)\right| = 0;\ s,\ t \in S$ implies $s = t$

then $\left|\bar{I}_n(s;t)\right|$ is a metric on S. The assertion follows for $\left|\bar{I}_n(s;t)\right|$. The result for $|I_n(s;t)|$ is obtained by showing that conditions (i) – (iv) above hold a.e., replacing E_* by $E_*^{U_{n-1}}$ and following the same steps. □

Proof of Theorem 2.9: The proof of properties (i)–(iii) is obtained precisely as in the proof of Theorem 2.8. Property (iv) is satisfied if and only if $\mathrm{f}_t\left(z_n \left|Z^{n-1}\right.\right) = \mathrm{f}_*\left(z_n \left|Z^{n-1}\right.\right)$ a.e. implies $t = *$ by Theorem 2.7.

Proof of Corollary 2.1: By Theorem 2.6 we have

$$I_n(r;t) - I_n(s;t) = I_n(r;s) \geq 0 \text{ a.s.}$$

and

$$\bar{I}_n(r;t) - \vec{I}_n(s;t) = \bar{I}_n(r;s) \geq 0$$

with equality if and only if $\mathrm{f}_s\left(z_n \left|Z^n\right.\right) = \mathrm{f}_r\left(z_n \left|Z^n\right.\right)$ a.e. The assertion follows. □

Proof of Corollary 2.2: The parameter s is closer to the true parameter than t in the metric $|I_n(s;t)|$ if and only if

$$|I_n(*;s)| < |I_n(*;t)| \text{ a.e.}$$

But, by Theorem 2.6

$$|I_n(*;s)| = I_n(*;s) \text{ a.e. for any } s \in S$$

Hence, s is closer to the true parameter than t if and only if

$$I_n(*;s) < I_n(*;t) \text{ a.e.}$$

or

$$I_n(*;t) - I_n(*;s) = I_n(s;t) > 0 \text{ a.e.}$$

To show that s is closer to the true parameter than t in the metric $|\bar{I}_n(s;t)|$, an identical procedure can be followed using $|\bar{I}_n(s;t)|$ instead of $I_n(s;t)$. □

Proof of Theorem 3.1: It follows from Eq. (3.3) that

$$(zI - A)\Sigma\left(z^{-1}I - A^T\right) + A\Sigma\left(z^{-1}I - A^T\right) + (zI - A)\Sigma A^T + \bar{K}C\Sigma A^T = BQB^T$$

where z is the Z transform variable and

$$\bar{K} = AK$$

We multiply both sides of this equation by $(zI - A)^{-1}$ on the left, and by $\left(z^{-1}I - A^T\right)^{-1}C^T$ on the right, and obtain

$$\begin{aligned}\Sigma C^T &+ (zI - A)^{-1}A\Sigma C^T + \Sigma A^T\left(z^{-1}I - A^T\right)^{-1}C^T \\ &+ (zI - A)^{-1}\bar{K}C\Sigma A^T\left(z^{-1}I - A^T\right)^{-1}C^T \\ &= (zI - A)^{-1}BQB^T\left(z^{-1}I - A^T\right)^{-1}C^T\end{aligned}$$

Since $A\Sigma C^T = \bar{K}R$, we readily find then that

$$\Sigma C^T + (zI - A)^{-1} \bar{K}R \left[I + \bar{K}^T \left(z^{-1}I - A^T\right)^{-1} C^T\right]$$
$$= (zI - A)^{-1} BQB^T \left(z^{-1}I - A^T\right)^{-1} C^T \tag{A.5}$$

Next, we consider the partial fraction expansion of the two sides of the last equation. Since the eigenvalues of A may consist of Jordan blocks of dimension greater than one, we consider the contribution of each Jordan block to the partial fraction expansion separately. It is well known that the terms in the partial fraction expansion of $(zI - A)^{-1}$ that correspond to the Jordan block of dimension r_i of the eigenvalue α_i are

$$\vartheta_{i,k,j}(z) = u_i^{(k)} v_i^{(j)T} (z - \alpha_i)^{-(r_i - j - k)}, \; j, k = 0, 1, \ldots, r_i - 1 \tag{A.6}$$

where $u_i^{(0)}$ and $v_i^{(0)T}$ are the eigenvector and the eigenrow that correspond to this Jordan block and $u_i^{(k)}$ and $v_i^{(k)T}$ are the corresponding k-th order pseudo-eigenvector and pseudo-eigenrows, respectively (Shaked and Dickson, 1977).

It follows from the asymptotic stability of the system Eq. (3.1) that the terms in the partial fraction expansion of $(zI - A)^{-1}\ BQB^T(z^{-1}I - A^T)^{-1}C^T$ that correspond to the eigenvalues of A are of the form $\vartheta_{i,k,j}(z)BQ\lim_{z\to\alpha_i} B^T(z^{-1}I - A^T)^{-1}C^T$. Obviously, there may be terms in the expansion that correspond to $(z - \alpha_i)^{-r_i}$. These terms result from different Jordan blocks that correspond to α_i and α_j where $\alpha_i = \alpha_j$. They include $\vartheta_{i,k,j}(z)BQ\lim_{z\to\alpha_i} B^T(z^{-1}I - A^T)^{-1}C^T$ for all $\alpha_i = \alpha_j$ and $r_i = r_j$ and also other dyadic terms that result from Jordan blocks of $\alpha_i = \alpha_j$, which are of dimension greater than r_i (i.e., $\vartheta_{j,k,r_j - r_i - k}(z)BQ\lim_{z\to\alpha_i} B^T(z^{-1}I - A^T)^{-1}C^T$, $k \geq r_j - r_i \geq 0$). Since the eigenvectors and pseudo-eigenvectors of A are all independent, it is possible, given the matrix coefficients of $(z - \alpha_i)^{-r_i}$, in the partial fraction expansion of $(zI - A)^{-1}BQB^T(z^{-1}I - A^T)^{-1}C^T$, to find the term that corresponds to $(z - \alpha_i)^{-r_i}$ and the eigenrow $v_i^{(0)T}$ of the specific Jordan block of α_i, namely $\vartheta_{0,0,0}(z)BQ\lim_{z\to\alpha_i} B^T(z^{-1}I - A^T)^{-1}C^T$.

The corresponding term in the partial fraction expansion of $(zI - A)^{-1}\bar{K}R[I + \bar{K}^T(z^{-1}I - A^T)^{-1}C^T]$ in Eq. (A.5) is clearly $\vartheta_{i,0,0}(z)\bar{K}R\lim_{z\to\alpha_i}F^T(z^{-1})$, where $F(z) = I + C(zI - A)^{-1}\bar{K}$ is the return difference matrix of the one step ahead stationary Kalman filter. It follows therefore from Eqs. (A.5) and (A.6) that

$$v_i^{(0)T}\bar{K}R\lim_{z\to\alpha_i}F^T\left(z^{-1}\right) = v_i^{(0)T}BQ\lim_{z\to\alpha_i}B^T\left(z^{-1}I - A^T\right)^{-1}C^T \tag{A.7}$$

By the Popov–Belevitch–Hautus test for reachability (Kailath, 1979, p. 135), $(A, \bar{K}R^{1/2})$ is reachable if and only if none of the eigenrows of A lie in the left annihilating subspace of $\bar{K}$. Since the zeros of $F^T(z^{-1})$ are the reciprocals of the poles of the Kalman filter, they lie outside the unit circle and $\lim_{z\to\alpha_i}F^T(z^{-1})$ is therefore nonsingular. The eigenrow $v_i^{(0)T}$ will lie in the annihilating subspace of $\bar{K}$ if and only if

$$v_i^{(0)T}BQ\lim_{z\to\alpha_i}B^T\left(z^{-1}I - A\right)^{-1}C^T = 0 \tag{A.8}$$

It follows from Eq. (A.8) that the pair $(A, \bar{K}R^{1/2})$ is not reachable if and only if one of the three cases holds:

(i) The matrix A is singular: The limit for $\alpha_i = 0$ will then yield the required zero in Eq. (A.8) independently of the specific structure of B and the eigenrows of A.
(ii) The pair (A, B) is nonreachable: In this case $v_i^{(0)T}B = 0$, which readily satisfies Eq. (A.8) independently of the specific structure of C.
(iii) The pair (A, B) is reachable and $\alpha_i \neq 0$ for each i, but Eq. (A.8) is satisfied for some i.

Since reachability of $(A, \bar{K}R^{1/2})$ implies that Π has a full rank (Anderson and Moore, 1979, p. 64) and dim z_n = rank Π, it follows that the state estimator Eq. (3.4) is minimal if and only if A is nonsingular, (A, B) is reachable, and none of the eigenrows of A satisfies Eq. (A.8). The assertion follows. □

Proof of Theorem 5.1: Since

$$E\left\{(x-\hat{x})^T(\hat{x}-\tilde{x})\right\}=0$$

we have

$$\begin{aligned}
&E\left\{(x-\tilde{x})^T(x-\tilde{x})\right\} \\
&\quad = E\{[(x-\hat{x})+(\hat{x}-\tilde{x})]^T[(x-\hat{x})+(\hat{x}-\tilde{x})]\} \\
&\quad = E\left\{(x-\hat{x})^T(x-\hat{x})\right\}E\left\{(\hat{x}-\tilde{x})^T(\hat{x}-\tilde{x})\right\} \qquad \text{(A.9)}
\end{aligned}$$

Let us denote by W the matrix whose columns are the normalized eigenvectors of $E\{\tilde{x}\tilde{x}^T\}$ corresponding to the eigenvalues arranged in descending order of magnitude. Let us write $W = [U \ \mathrm{V}]$ where U has r columns. Then

$$W^T\tilde{x}=\begin{bmatrix}\tilde{x}^r\\0\end{bmatrix} \quad \text{a.s.}$$

where $\tilde{x}^r$ is a vector of dimension r, yielding

$$\tilde{x}=W\begin{bmatrix}\tilde{x}^r\\0\end{bmatrix}=U\tilde{x}^r \quad \text{a.s.} \qquad \text{(A.10)}$$

Since W is orthogonal

$$WW^T=UU^T+VV^T=I$$

Hence

$$\begin{aligned}
&E\left\{(\hat{x}-\tilde{x})^T(\hat{x}-\tilde{x})\right\}=E\left\{(\hat{x}-U\tilde{x})^T(\hat{x}-U\tilde{x}^r)\right\} \\
&\quad = E\left\{(\hat{x}-U\tilde{x}^r)^T WW^T(\hat{x}-U\tilde{x}^r)\right\}
\end{aligned}$$

yielding

$$\begin{aligned}
&E\left\{(\hat{x}-\tilde{x})^T(\hat{x}-\tilde{x})\right\}=E\left\{(\hat{x}-U\tilde{x})^T UU^T(\hat{x}-U\tilde{x}^r)\right\} \\
&\quad = E\left\{(\hat{x}-U\tilde{x}^r)^T VV^T(\hat{x}-U\tilde{x}^r)\right\} \qquad \text{(A.11)}
\end{aligned}$$

which implies

$$\begin{aligned}
&E\left\{(\hat{x}-\tilde{x})^T(\hat{x}-\tilde{x})\right\}\geq E\left\{(\hat{x}-U\tilde{x}^r)^T VV^T(\hat{x}-U\tilde{x}^r)\right\} \\
&\quad = E\left\{\hat{x}^T VV^T\hat{x}\right\} \qquad \text{(A.12)}
\end{aligned}$$

Noting that

$$VV^T = I - UU^T = \left(I - UU^T\right)\left(I - UU^T\right)$$

we see that

$$\begin{aligned} E\left\{\hat{x}^T VV^T \hat{x}\right\} &= E\left\{\hat{x}^T \left(I - UU^T\right)\left(I - UU^T\right)\hat{x}\right\} \\ &= \operatorname{tr}\left\{\left(I - UU^T\right)\Pi\left(I - UU^T\right)\right\} \\ &= \operatorname{tr}\left\{\left(\Pi^{1/2} - UU^T\Pi^{1/2}\right)\left(\Pi^{1/2} - UU^T\Pi^{1/2}\right)\right\} \end{aligned}$$

Since, clearly, $UU^T\Pi^{1/2}$ has, at most, rank r, it follows from a well-known result, e.g., (Rao, 1973, p. 631), that

$$E\left\{\hat{x}^T VV^T \hat{x}\right\} \geq \sum_{i=r+1}^{n} \lambda_i \tag{A.13}$$

with equality if and only if

$$T^T UU^T \Pi^{1/2} T = \Phi \tag{A.14}$$

where

$$\Phi = \operatorname{diag}\left(\lambda_i^{1/2}, \ldots, \lambda_r^{1/2}, 0, \ldots, 0\right)$$

The asserted inequality Eq. (5.1) now follows from Eqs. (A.9), (A.12) and (A.13). To see that Eq. (5.4) is necessary and sufficient for equality in Eq. (5.1), we first note that the second term on the right-hand side of Eq. (A.11) is minimal if and only if Eq. (A.14) holds, yielding, by Eq. (A.13),

$$T^T UU^T T \Lambda^{t/2} = \Phi$$

or

$$TUU^T T = \begin{bmatrix} I & 0 \\ 0 & 0 \end{bmatrix}$$

yielding

$$UU^T = T\begin{bmatrix} I & 0 \\ 0 & 0 \end{bmatrix}T^T = T\begin{bmatrix} 1 \\ 0 \end{bmatrix}\left[I \; 0\right]T^T = T^r T^{r^T} \tag{A.15}$$

The first term on the right-hand side of Eq. (A.11) vanishes if and only if

$$U^T\left(\hat{x}-U\tilde{x}^r\right)=0 \quad \text{a.s.}$$

or

$$\tilde{x}^r=U^T\hat{x} \quad \text{a.s.}$$

yielding, by Eq. (A.10)

$$\tilde{x}=UU^T\hat{x} \quad \text{a.s.} \tag{A.16}$$

It follows from Eqs. (A.15) and (A.16) that Eq. (5.4) is necessary and sufficient for equality in Eq. (5.1). The proof is thus completed. □

Proof of Theorem 5.2: The ideal reduced-order estimator is realizable if and only if x_k^r and x_{k+1}^r, defined by Eqs. (5.15) and (5.16) are related by

$$x_{k+1}^r=A_k^r x_k^r+K_k^r\left(y_k-C_k^r x_k^r\right) \quad \text{a.s.} \tag{A.17}$$

where A_k^r, C_k^r and K_k^r are defined by Eqs. (5.25), (5.26) and (5.27), respectively, i.e., if and only if

$$T_{k+1}^{rT}\hat{x}_{k+1}^r=A_k^r T_k^{rT}\hat{x}_k^r+K_k^r\left(y_k-C_k^r T_k^{rT}\hat{x}_k^r\right) \quad \text{a.s.}$$

But, from the Kalman filter model (Kalman, 1960), we have

$$T_{k+1}^{rT}\hat{x}_{k+1}=T_{k+1}^{rT}A_k\hat{x}_k+T_{k+1}^{rT}K_k\left(y_k-C_k\hat{x}_k\right) \quad \text{a.s.}$$

The ideal reduced-order estimator is realizable, then, if and only if the equality

$$\begin{aligned}\left(A_k^r T_k^{rT}-K_k^r C_k^r T_k^{rT}\right)\hat{x}_k+K_k^r y_k&=T_{k+1}^{rT}\left(A_k-K_k C_k\right)\hat{x}_k\\&\quad+T_{k+1}^{rT}K_k y_k \quad \text{a.s.}\end{aligned} \tag{A.18}$$

holds for each k. Rearranging Eq. (A.18), we obtain

$$\left[\left(A_k^r - K_k^r C_k^r\right) T_k^{rT} - T_{k+1}^{rT} A_k + K_k^r C_k\right] \hat{x}_k = \left(T_{k+1}^{rT} K_k - K_k^r\right) \upsilon_k \quad \text{a.s.} \tag{A.19}$$

Since $\hat{x}_k$ and υ_k are uncorrelated, Eq. (A.19) holds if and only if both sides are, a.s., zero. The right-hand side of Eq. (A.19) is, a.s., zero, if and only if

$$\left(T_{k+1}^r K_k - K_k^r\right)\left(C_k P_k C_k^T + R_k\right)\left(T_{k+1}^r K_k - K_k^r\right)^T = 0 \tag{A.20}$$

Since R_k is assumed to have a full rank, $\left(C_k P_k C_k^T + R_k\right)$ has a full rank and Eq. (A.20) holds if and only if

$$K_k^r = T_{k+1}^{rT} K_k \tag{A.21}$$

Substituting Eqs. (5.25), (5.26) and (A.21) into Eq. (A.19), we get

$$\left[T_{k+1}^{rT}\left(A_k - K_k C_k\right) T_k^r T_k^{rT} - T_{k+1}^{rT}\left(A_k - K_k C_k\right)\right] \hat{x}_k = 0 \quad \text{a.s.}$$

yielding

$$T_{k+1}^{rT} \Gamma_k \left(I - T_k^{rT}\right) \hat{x}_k = 0 \quad \text{a.s.}$$

The latter holds, a.s., if and only if

$$T_{k+1}^{rT} \Gamma_k \left(I - T_k^r T_k^{rT}\right) \Pi_k \left(I - T_k^r T_k^{rT}\right)^T \Gamma_k^T T_{k+1}^r = 0$$

or, substituting Eq. (5.23)

$$T_{k+1}^{rT} \Gamma_k \left(I - T_k^r T_k^{rT}\right) T_k \Lambda_k T_k^T \left(I - T_k^r T_k^{rT}\right)^T \Gamma_k^T T_{k+1}^r = 0$$

yiolding

$$T_{k+1}^{rT} \Gamma_k T_k \left(I - \begin{bmatrix} I_r & 0 \\ 0 & 0 \end{bmatrix}\right) \Lambda_k \left(I - \begin{bmatrix} I_r & 0 \\ 0 & 0 \end{bmatrix}\right) T_k^T \Gamma_k^T T_{k+1}^r = 0$$

or

$$T_{k+1}^{rT}\Gamma_k T_k \bar{\Lambda}_k^r T_k^T \Gamma_k T_{k+1}^r = 0 \tag{A.22}$$

But Eq. (A.22) holds if and only if

$$[I_r \ 0]\, T_{k+1}^T \Gamma_k T_k \bar{\Lambda}_k^{r1/2} = 0 \tag{A.23}$$

where

$$\bar{\Lambda}_k^{r1/2} = \begin{bmatrix} 0 & & & & & \\ & \ddots & & & & 0 \\ & & 0 & & & \\ & & & \lambda_{k,r+1}^{1/2} & & \\ & & & & \ddots & \\ & 0 & & & & \lambda_{k,n}^{1/2} \end{bmatrix}$$

Since, as we can readily verify

$$[I_r \ 0]\, T_{k+1}^T \Gamma_k T_k \bar{\Lambda}_k^{r1/2} = \left[0 \ \Theta_k \Lambda_k^{-1}\right] \tag{A.24}$$

we see that Eq. (A.23) holds and, hence, the left-hand side of Eq. (A.19) is zero, a.s., if and only if

$$\Theta_k \Lambda_k^{-1} = 0$$

We have shown, then, that Eq. (A.17) holds if and only if Eqs. (A.19) and (A.24) hold simultaneously. The proof is thus completed. □

Proof of Corollary 5.1: If $l = 0$, we have

$$\bar{\Lambda}_k^r = 0 \tag{A.25}$$

yielding

$$P_k^r = P_k + T_k \bar{\Lambda}_k^r T_k^{rT} = P_k$$

It follows that

$$K_k^r = T_{k+1}^{rT} A_k P_k C_k^T \left[C_k P_k C_k^T + R_k \right]^{-1} = T_{k+1}^{rT} K_k$$

hence, condition (i) of Theorem 5.2 is satisfied. Furthermore, Eq. (A.25) implies

$$\Lambda_k^{-r} = 0$$

hence, condition (ii) of Theorem 5.2 is also satisfied. For $l \leq 0$, the ideal estimator is realizable then by Theorem 5.2. For $l > 0$, condition (ii) of Theorem 5.2 is satisfied if and only if Eq. (5.37) is satisfied. The assertion follows. □

Proof of Theorem 9.1: Since the set $j \in k;\ j \neq k$ is finite, condition (c9.1) implies

$$\lim_{n \to \infty} \sup_j \left\{ h_k^j (Z^n);\ j \in K;\ j \neq k \right\} = 0 \text{ a.e.}$$

Hence

$$\lim_{n \to \infty} \sup_j \left\{ h_k^j (Z^n);\ j \in K \right\} = h_k^j (Z^n) = 1 \text{ a.e.}$$

or

$$\lim_{n \to \infty} \hat{k} (Z^n) = k \text{ a.e., as asserted.}$$

□

Proof of Theorem 9.2: By Eq. (2.19), we have, for each $j \in k$

$$\mathrm{f}^b (j \,|Z^n) \leq \frac{\mathrm{f}_0^b (j) \, \mathrm{f}_j (Z^n)}{\mathrm{f}_0^b (k) \, \mathrm{f}_k (Z^n)} = \frac{\mathrm{f}_0^b (j)}{\mathrm{f}_0^b (k)} h_k^j (Z^n)$$

By condition (c9.1), for each $j \in k;\ j \neq k$, we have

$$\lim_{n \to \infty} \mathrm{f}^b (j \,|Z^n) \leq \frac{\mathrm{f}_0^b (j)}{\mathrm{f}_0^b (k)} \lim_{n \to \infty} h_k^j (Z^n) = 0 \quad \text{a.e.}$$

implying

$$\lim_{n\to\infty} \mathrm{f}^b(j\,|Z^n) = 0 \text{ a.e. for each}$$
$$j \in K; \quad j \neq k$$

But since

$$\sum_{j=0}^{p} \mathrm{f}^b(j\,|Z^n) = 1$$

we have

$$\lim_{n\to\infty} \mathrm{f}^b(k\,|Z^n) = 1 \text{ a.e.}$$

yielding the assertion. □

Proof of Theorem 9.3: By Eqs. (2.20) and (9.2), we have

$$\lim_{n\to\infty} \hat{s}_n = \sum_{j=0}^{p} s_j \lim_{n\to\infty} \mathrm{f}^b(j\,|Z^n) = s_k \text{ a.e.}$$

The assertion follows. □

Proof of Theorem 9.4: We follow, in part, a previous analysis (Liporace, 1971), addressing the case of independent and identically distributed observations. Consider the norm

$$\begin{aligned}
N_n &= E_*\left\{(\hat{s}_n - s_k)^T(\hat{s}_n - s_k)\right\} \\
&= E_*\left\{\sum_{j=0}^{p}(s_j - s_k)^T \mathrm{f}^b(s_j\,|Z^n)\sum_{i=0}^{p}(s_i - s_k)^T \mathrm{f}^b(s_i\,|Z^n)\right\} \\
&= \sum_{j=0}^{p}\sum_{i=0}^{p}(s_j - s_k)^T(s_i - s_k)\,E_*\left\{\mathrm{f}^b(s_j\,|Z^n)\,\mathrm{f}^b(s_i\,|Z^n)\right\} \\
&\le p^2R^2E_*\left\{\mathrm{f}^b(s_j\,|Z^n)\right\} \quad \text{for some } j \in K;\ j \neq k
\end{aligned}$$

where

$$R^2 = \max\left\{(s_j - s_k)^T(s_j - s_k);\ j \in K;\ j \neq k\right\}$$

since, obviously

$$E_*\left\{\mathrm{f}^b\left(s_j\,|Z^n\right)\right\}\geq E_*\left\{\mathrm{f}^b\left(s_j\,|Z^n\right)\mathrm{f}\left(s_i\,|Z^n\right)\right\}$$

By Eq. (2.19), we have for each $j\in K;\ j\neq k$

$$\mathrm{f}^b\left(s_j\,|Z^n\right)\leq\frac{\mathrm{f}_0^b\left(s_j\right)\mathrm{f}_j\left(Z^n\right)}{\mathrm{f}_0^b\left(s_k\right)\mathrm{f}_k\left(Z^n\right)}=\frac{\mathrm{f}_0^b\left(s_j\right)}{\mathrm{f}_0^b\left(s_k\right)}h_k^j\left(Z^n\right)$$

By condition (c9.1), we have for each $j\in K;\ j\neq k$

$$\lim_{n\to\infty}\mathrm{f}^b\left(s_j\,|Z^n\right)\leq\frac{\mathrm{f}_0^b\left(s_j\right)}{\mathrm{f}_0^b\left(s_k\right)}\lim_{n\to\infty}h_k^j\left(Z^n\right)=0\text{ a.e.}$$

Hence

$$\lim_{n\to\infty}\mathrm{f}^b\left(s_j\,|Z^n\right)=0\text{ a.e.}$$

Now since

$$\mathrm{f}^b\left(s_j\,|Z^n\right)\leq 1$$

we have by the dominated convergence theorem (Theorem 2.3)

$$\lim_{n\to\infty}E_*\mathrm{f}^b\left(s_j\,|Z^n\right)=E_*\lim_{n\to\infty}\mathrm{f}^b\left(s_j\,|Z^n\right)=0$$

and thus

$$\lim_{n\to\infty}N_n\leq p^2R^2\lim_{n\to\infty}E_*\mathrm{f}^b\left(s_j\,|Z^n\right)=0$$

yielding finally

$$\lim_{n\to\infty}N_n=0$$

The assertion follows. □

Proof of Lemma 9.1: We have noted (see Section 2.4) that for each $j\subset K$, the sequence of likelihood ratios $(h_*^j(Z^n))$ is a U_n-martingale according to the measure P_*. Furthermore

$$E_*h_*^j\left(Z^n\right)=0\text{ a.e.}$$

It follows from the martingale convergence theorem (Theorem 2.4) that the sequence $(h_*^j(Z^n))$ converges to a finite limit. Thus, the sequence $(\log h_*^j(Z^n))$ converges to some $a < \infty$. We have

$$\log h_j^* (Z^n) = Y_n (*; j) + V_n (*; j) \tag{A.26}$$

Suppose that

$$\lim_{n\to\infty} \log h_*^j (Z^n) = a > -\infty \text{ a.e.}$$

or

$$\lim_{n\to\infty} \log h_j^* (Z^n) = a < \infty \text{ a.e.}$$

Then by condition (c9.3) and by Eq. (A.26), we have

$$\lim_{n\to\infty} V_n (*; j) = -\infty \text{ a.e.}$$

Contradicting condition (c9.2). Hence, we have

$$\lim_{n\to\infty} \log h_j^* (Z^n) = \infty \text{ a.e.}$$

or

$$\lim_{n\to\infty} \log h_*^j (Z^n) = -\infty \text{ a.e.}$$

yielding

$$\lim_{n\to\infty} h_*^j (Z^n) = 0 \text{ a.e.}$$

The assertion follows.

□

Theorem 9.5 follows directly from Lemma 9.1 and Theorems 9.1–9.4. □

Proof of Theorem 9.6: By Theorem 2.6 we have

$$I_n(k;j) \geq 0 \text{ a.e. for all } n \geq 0$$

Thus, condition (c9.4) implies condition (c9.3). The assertion follows from Theorem 9.5. □

Proof of Lemma 9.2: We have by Eq. (2.21) for each $\omega \in \Omega$

$$V_n(k;j,\omega) = J_1(k;j,\omega) + V_{n-1}(k;j,T\omega)$$

where T is a measure preserving transformation. It follows that the event is invariant. Thus, either

$$P\{\limsup V_n(k;j) < \infty\} = 0$$

or

$$P\{\limsup V_n(k;j) < \infty\} = 1$$

Obviously, we have that if

$$\lim\ \sup V_n(k;j) < \infty$$

then

$$\limsup \frac{V_n(k;j)}{\sqrt{n}} < \infty$$

But, by Theorem 2.6

$$P\left\{\limsup \frac{V_n(k;j)}{\sqrt{n}} < \infty\right\} < 1$$

Hence

$$P\{\limsup V_n(k;j) < \infty\} < 1$$

yielding

$$P\{\limsup V_n(k;j) < \infty\} = 0$$

Thus

$$\limsup V_n(k;j) = \infty \quad \text{a.e.}$$

The assertion follows. □

Proof of Lemma 9.3: Let

$$R_a(k;j) \equiv V_{n_a}(k;j) - a$$

and

$$V_n^a(k;j) \equiv V_{\min(n,n_a)}(k;j)$$

Note that

$$V_n^a(k;j) \leq a + R(k;j)$$

Since $Y_n(k;j)$ is a U_n-martingale, so is $(Y_n^a(k;j))$. Obviously, we have

$$E_*\left\{V_n^{a^+}(k;j)\right\} \leq a + E_* R_a(k;j)$$

Hence, under Eq. (9.2)

$$E_*\left\{V_n^{a^+}(k;j)\right\} \leq \infty$$

It follows from Theorem 2.4 that the sequence $(V_n^a(k;j))$ converges to a finite limit. Let

$$A_a \equiv \{\omega \in \Omega : \sup V_n(k;j,\omega) < a\}$$

and

$$A \equiv \cup_{a=1}^{\infty} A_a$$

If $\omega \in A$, then $\omega \in A_a$ for some a. Then,

$$V_n(k;j,\omega) = V_n^{a_0}(k;j,\omega) \text{ for all } n$$

and then

$$\lim_{n\to\infty} V_n\left(k;j,\omega\right) = \lim_{n\to\infty} V_n^{a_0}\left(k;j,\omega\right) \text{ is finite.}$$

If $\omega \notin A$, then

$$\limsup V_n\left(k;j,\omega\right) = \infty.$$

The assertion follows. □

Proof of Theorem 9.7: Since the sequence $\left(h_k^j\left(Z^n\right)\right)$ is a (U_n, P_k)-martingale and since

$$E_k h_k^j\left(Z^n\right) = 1$$

it follows from Theorem 2.4 that $(h_k^j(Z^n))$ converges a.e. P_k to a finite random variable.

Since P_* is absolutely continuous with respect to P_k, then $(h_k^j(Z^n))$ converges to a finite random variable P_*.

The remainder of the proof is identical to the proof of Lemma 9.1, and the convergence of the estimates follows from the convergence of the likelihood ratios by Theorems 9.1 through 9.4. The assertion follows. □

Proof of Theorem 9.8: For each $j \in K$, the likelihood ratio sequence $\left(\frac{\mathrm{f}_k(Z^n)}{\mathrm{f}_j(Z^n)}\right)$ is a martingale according to the measure P_j (Doob, 1953, p. 93). In addition, we have

$$E_j\left\{\frac{\mathrm{f}_k\left(Z^n\right)}{\mathrm{f}_j\left(Z^n\right)}\right\} = 1$$

By the martingale convergence theorem, we then have

$$\lim_{n\to\infty} \frac{\mathrm{f}_k\left(Z^n\right)}{\mathrm{f}_j\left(Z^n\right)} = \text{finite r.v. a.e. } \mathrm{P}_j$$

(where r.v. denotes random variable.)

But under condition (c9.1)

$$\lim_{n\to\infty} \frac{\mathrm{f}_k(Z^n)}{\mathrm{f}_j(Z^n)} = \infty \quad \text{a.e. } \mathrm{P}_*$$

So we have

$$P_j\left\{\lim_{n\to\infty} \frac{\mathrm{f}_k(Z^n)}{\mathrm{f}_j(Z^n)} = \text{finite r.v.}\right\} = 1$$

and also

$$P_*\left\{\lim_{n\to\infty} \frac{\mathrm{f}_k(Z^n)}{\mathrm{f}_j(Z^n)} = \text{finite r.v.}\right\} = 0$$

Hence, under condition (c9.1), the measure P_j; $j \in K$; $j \neq k$ is singular with respect to the measure P_*. The assertion follows. □

Proof of Theorem 9.9: We have by condition (c9.1) for each $j \in K$: $j \neq k$.

$$\lim_{n\to\infty} h_j^k(Z^n) = \infty \quad \text{a.e.}$$

and by Eq. (9.2)

$$\lim_{n\to\infty} \frac{\mathrm{f}^b(k\,|Z^n)}{\mathrm{f}^b(j\,|Z^n)} = \infty \quad \text{a.e.}$$

Since both sequences are non-negative, we have by Fatou's lemma (Theorem 2.2)

$$\lim_{n\to\infty} E_* h_j^k(Z^n) \geq \lim_{n\to\infty} \inf E_* h_j^k(Z^n) \geq E_* \lim_{n\to\infty} \inf h_j^k(Z^n) = \infty$$

and

$$\lim_{n\to\infty} E_* \frac{\mathrm{f}^b(k\,|Z^n)}{\mathrm{f}^b(j\,|Z^n)} \geq \lim_{n\to\infty} \inf E_* \frac{\mathrm{f}^b(k\,|Z^n)}{\mathrm{f}^b(j\,|Z^n)}$$

$$\geq E_* \lim_{n\to\infty} \inf \frac{\mathrm{f}^b(k\,|Z^n)}{\mathrm{f}^b(j\,|Z^n)} = \infty$$

The assertion follows. □

Proof of Theorem 9.10:

$$\begin{aligned}\bar{I}_n(k;j) &= E_* \log \frac{\mathrm{f}_k\left(z_n \left| Z^{n-1}\right.\right)}{\mathrm{f}_j\left(z_n \left| Z^{n-1}\right.\right)} \\ &= E_* \log \frac{\mathrm{f}_k(z_n)}{\mathrm{f}_j(z_n)} - E_* \log \frac{\mathrm{f}_k\left(Z^{n-1}\right)}{\mathrm{f}_j\left(Z^{n-1}\right)}\end{aligned}$$

By condition (c9.5), we then have

$$E_* \log \frac{\mathrm{f}_k(z_n)}{\mathrm{f}_j(z_n)} - E_* \log \frac{\mathrm{f}_k\left(Z^{n-1}\right)}{\mathrm{f}_j\left(Z^{n-1}\right)} \geq \alpha_j$$

yielding

$$E_* \log \frac{\mathrm{f}_k(z_n)}{\mathrm{f}_j(z_n)} \geq a_j + (n - N_j)\,\alpha_j \quad \text{for all } n \geq N_j \tag{A.27}$$

where

$$a_j \equiv E_* \log \frac{\mathrm{f}_k\left(z_{N_j} \left| Z^{N_j - 1}\right.\right)}{\mathrm{f}_j\left(z_{N_j} \left| Z^{N_j - 1}\right.\right)} = \bar{I}_{N_j}(k;j) \geq \alpha_j \tag{A.28}$$

Since log(∗) is a concave function, we have by Jensen's inequality (Theorem 2.1)

$$\log E_* \frac{\mathrm{f}_k(Z^n)}{\mathrm{f}_j(Z^n)} \geq E_* \log \frac{\mathrm{f}_k(Z^n)}{\mathrm{f}_j(Z^n)} \tag{A.29}$$

Equations (A.27), (A.28), (A.29) together yield

$$E_* h_j^k(Z^n) = E_* \frac{\mathrm{f}_k(Z^n)}{\mathrm{f}_j(Z^n)} \geq e^{a_j} e^{(n-N_j)\alpha_j} \geq e^{(n-N_j+1)\alpha_j} \tag{A.30}$$

for all $n \geq N_j$, for each $j \in K;\ j \neq k$.

Hence, for each $j \in K;\ j \neq k$, the likelihood ratio $h_j^k(Z^n)$ tends in the mean to infinity faster than an exponential with a rate of α_j.

By Eq. (2.19), we have for each $j \in K$

$$\frac{\mathrm{f}^b(k\,|Z^n)}{\mathrm{f}^b(j\,|Z^n)} = \frac{\mathrm{f}^b(k)}{\mathrm{f}^b(j)}\frac{\mathrm{f}_k(Z^n)}{\mathrm{f}_j(Z^n)}$$

Thus, by Eq. (A.30), for each $j \in K;\ j \neq k$

$$E_*\frac{\mathrm{f}^b(k\,|Z^n)}{\mathrm{f}^b(j\,|Z^n)} = \frac{\mathrm{f}^b(k)}{\mathrm{f}^b(j)}E_*\frac{\mathrm{f}_k(Z^n)}{\mathrm{f}_j(Z^n)}$$

$$\geq \frac{\mathrm{f}^b(k)}{\mathrm{f}^b(j)}e^{(n-N_j+1)\alpha_j} \quad \text{for all } n \geq N_j \tag{A.31}$$

Hence, for each $j \in K;\ j \neq k$, the *a posteriori* probability ratio $\frac{\mathrm{f}^b(k|Z^n)}{\mathrm{f}^b(j|Z^n)}$ tends in the mean to infinity faster than an exponential with a rate of α_j.

Finally, taking $N = \max\{N_j;\ j \in K;\ j \neq k\}$ and $\alpha = \min\{\alpha_j;\ j \in K;\ j \neq k\}$ implies that the sequences $(h_j^k(Z^n))$ and $\left(\frac{\mathrm{f}^b(k|Z^n)}{\mathrm{f}^b(j|Z^n)}\right)$ converge in L_1 to infinity faster than an exponential with a rate of α for all $n \geq N$, as asserted. □

Proof of Theorem 10.1: Since both (z_n) and $(\hat{z}_{s,n})$ are obtained by linear operations on zero-mean Gaussian sequences, they are zero-mean Gaussian, and so is each of the sequences $(z_n - \hat{z}_{s,n});\ s \in T;\ n \geq 0$. The existence of Γ_s^* implies stationarity of $(z_n - \hat{z}_{s,n})$. It follows that $(z_n - \hat{z}_{s,n})$ is ergodic if and only if

$$\lim_{n\to\infty}\frac{1}{n+1}\sum_{k=0}^{n}|R(K)|^2 = 0 \tag{A.32}$$

where $|R(k)|$ denotes the determinant of the matrix

$$\begin{aligned} R(k) &= E\left\{(z_n - \hat{z}_{s,n})(z_{n+k} - \hat{z}_{s,n+k})^T\right\} \\ &= \begin{cases} H_s^t\Psi_s^t H_s^{t^T} + R_t, & k = 0 \\ H_s^t\Psi_s^t H_s^{t^T}\left(F_s^t\right)^k, & k > 0 \end{cases} \end{aligned} \tag{A.33}$$

We have, for any $k > 0$

$$|R(k)| = \left|H_s^t\Psi_s^t H_s^{t^T}\right|\left|F_s^t\right|^k$$

Since all eigenvalues of F_s^t are inside the unit circle, then $\left|F_s^t\right| < 1$. Hence

$$\lim_{n\to\infty}\sum_{k=1}^{n}\left|R\left(k\right)\right|^2 = \left|H_s^t\Psi_s^t H_s^{t^T}\right|^2 \lim_{n\to\infty}\sum_{k=1}^{n}\left|F_s^t\right|^{2k} = \frac{\left|H_s^t\Psi_s^t H_s^{t^T}\right|^2}{1-\left|F_s^t\right|^2} < \infty \tag{A.34}$$

yielding

$$\lim_{n\to\infty}\frac{1}{n+1}\sum_{j=0}^{n}\left|R\left(k\right)\right|^2 = 0$$

The assertion follows. □

Proof of Theorem 10.2:

$$\log h_t^s\left(Z^n\right) = \sum_{m=1}^{n}\log h_t^s\left(z_m\left|Z^{m-1}\right.\right)$$

we have

$$\begin{aligned}\log h_t^s\left(z_n\left|Z^{n-1}\right.\right) &= \frac{1}{2}\log\frac{\Sigma_t}{\Sigma_s}\\ &\quad -\frac{1}{2}\left(z_n-\hat{z}_{s,n}\right)^T\Sigma_s^{-1}\left(z_n-\hat{z}_{s,n}\right)\\ &\quad +\frac{1}{2}\left(z_n-\hat{z}_{t,n}\right)^T\Sigma_t^{-1}\left(z_n-\hat{z}_{t,n}\right)\end{aligned} \tag{A.35}$$

Since for each $s \in T$, the residuals $\left(z_n - \hat{z}_{s,n}\right)$ are ergodic, it follows from the ergodic theorem, e.g., (Doob, 1953)

$$\begin{aligned}\lim_{n\to\infty}\frac{1}{n}\sum_{m=1}^{n}\log h_t^s\left(z_m\left|Z^{m-1}\right.\right) &= E_*\log h_t^s\left(z_m\left|Z^{m-1}\right.\right)\\ &= I_m(s;t) = \frac{1}{2}\left(L_t^* - L_s^*\right)\end{aligned} \tag{A.36}$$

Now if

$$L_t^* < L_s^* \tag{A.37}$$

then, obviously

$$\lim_{n\to\infty}\sum_{m=1}^{n}\log h_t^s\left(z_m\left|Z^{m-1}\right.\right)=\lim_{n\to\infty}\log h_t^s\left(Z^n\right)=-\infty\quad\text{a.s.}$$

yielding

$$\lim_{n\to\infty} h_t^s\left(Z^n\right)=0$$

which contradicts Eq. (10.17). Thus, Eq. (10.17) implies (10.19). The assertion follows. □

Proof of Corollary 10.1: By Eq. (2.27), we have

$$I\left(*;j\right)\geq 0 \text{ for each } j\in k$$

Hence

$$L_j^*-L_*^*>0$$

if

$$L_j^*\neq L_*^* \tag{A.38}$$

The assertion now follows from Theorem 10.2. □

Proof of Theorem 10.4: By Theorem 10.2, condition (c10.3) implies condition (c9.1). The L_1 convergence of both sequences follows from Theorem 9.9. Now let

$$\alpha_j\equiv\frac{1}{2}\left(L_j^*-L_k^*\right)\text{ for each } j\in K;\ j\neq k$$

then following the proof of Theorem 9.10, we get by Eqs. (A.29) and (A.30) for each $j\in K;\ j\neq k$

$$E\left\{h_j^k\left(Z^n\right)\right\}\geq e^{(n+1)\alpha_j} \tag{A.39}$$

and

$$E\left\{\frac{\mathrm{f}^b\left(k\,|Z^n\right)}{\mathrm{f}^b\left(j\,|Z^n\right)}\right\} \geq \frac{\mathrm{f}^b\left(k\right)}{\mathrm{f}^b\left(j\right)}e^{(n+1)\alpha_j} \tag{A.40}$$

The assertion follows. □

Proof of Theorem 10.5: By Theorem 10.2, condition (c10.3) implies condition (c9.1). The L_1 convergence of both sequences $(h_j^k(Z^n))$ and $\frac{\mathrm{f}^b(k|Z^n)}{\mathrm{f}^b(j|Z^n)}$ follows from Theorem 9.10.

Now let

$$\alpha_j \equiv \frac{1}{2}\left[L_j^* - L_k^*\right] \text{ for each } j \in K;\ j \neq k$$

then employing Eqs. (4.11) and (4.12), we obtain for each $j \in K$; $j \neq k$

$$E\left\{h_j^k\left(Z^n\right)\right\} \geq e^{(n+1)\alpha_j}$$

and

$$E\frac{\mathrm{f}^b\left(k\,|Z^n\right)}{\mathrm{f}^b\left(j\,|Z^n\right)} \geq \frac{\mathrm{f}^b\left(k\right)}{\mathrm{f}^b\left(j\right)}e^{(n+1)\alpha_j}, \text{ as asserted.}$$

□

Proof of Theorem 12.1: It is well known that when the system is controllable by the input noise and observable, the state estimation error covariance has a nonsingular limit value. It follows that the output estimation error covariance $\Sigma_n(p, s)$ has a nonsingular limit value $\Sigma(p, s)$ if $H(p)$ has independent rows, which is the normal case. It is also well known that an equation of the type (A2) (a Lyapunov equation specified in the Appendix) has a limit solution for $\psi_n(p, s)$ if $\bar{F}(p, s)$ has all its eigenvalues inside the unit circle. But due to the structure of $\bar{F}(p, s)$, its eigenvalues are the eigenvalues of $F[p, s(p)]$, together with the eigenvalues of $F(p, s)[I - K(p, s)H(p)]$. Since s(p) is stabilizing, $F[p, s(p)]$ has all its eigenvalues inside the unit circle. Since the system $[F(p, s), G(p), H(p)]$ is controllable and observable, the matrix $F(p,s)[I - K(p,s)H(p)]$ has all its eigenvalues inside the unit circle (see, e.g., (Schweppe, 1973, p. 144)). It follows that all the eigenvalues of $\bar{F}(p,s)$ are inside the unit circle and, consequently, Eq. (A2) has a limit solution for ψ_n. In addition, since the system Eq. (A3) is stable, $\bar{X}_n(p,s)$ has a finite limit value. It follows that $\Gamma_n(p,s)$ in Eq. (Al) has a finite limit value $\Gamma(p,s)$. We have thus shown that $\Sigma_n(p,s)$ and $\Gamma_n(p,s)$ have limit values $\Sigma(p,s)$ and $\Gamma(p,s)$. The assertion readily follows from Eq. (12.3). $\square$

References

Akaike, H. (1972). Information theory and an extension of the maximum likelihood principle. *Proc. 2nd Int Symp Informat Theor.* Budapest.

Akaike, H. (1974a). A new look at the statistical system identification. *IEEE Trans Automat Contr* **19**(6):716–723.

Akaike, H. (1974b). Stochastic theory of minimal realization. *IEEE Trans Automat Contr* **19**:667–674.

Anderson, B. D. O. (1969). The inverse problem of stationary covariance generation. *J Stat Phys* **1**(1):133–147.

Anderson, B. D. O. and Moore, J. B. (1979). *Optimal Filtering.* Prentice-Hall, Englewood Cliffs, NJ.

Armutlu, K., Karabudak, R. and Nurlu, G. (2001). Physiotherapy approaches in the treatment of ataxic multiple sclerosis: A pilot study. *Neurorehabil Neural Repair* **15**:203–211.

Arumagan, M. and Ramamoorty, M. (1973). A method for simplifying large dynamic systems. *Int J Contr* **17**:1129–1135.

Aston-Jones, G., Rajkowski, J., Kubiak P. and Alexinsky, T. (1994). Locus coeruleus neurons in monkey are selectively activated by attended cues in a vigilance task. *J Neurosci* **14**:4467–4480.

Astrom, K. J. (1970). *Introduction to Stochastic Control Theory.* Academic Press, New York.

Athans, M. (1971). The role and use of the stochastic linear-quadratic-Gaussian problem in control system design. *IEEE Trans Automat Contr* **16**(6):529–552.

Athans, M., Ferki, S. and Pascoal, A. (2006). Issues on robust adaptive feedback control. *16th IFAC World Congress.* Prague Czech Republic, July 2005.

Athans, M., Castanon, D., Dunn, K. P., Green, C., Wing, H. L., Sandell, N. R. and Willsky, A. (1977). The stochastic control of the F-8C aircraft using a multiple model adaptive control (MMAC) method-part I; equilibrium flight. *IEEE Trans Automat Contr* **22**(5):768–780.

Azulay, J. P., Mesure, S., Amblard, B., Blin, O., Sangla, I. and Pouget, J. (1999). Visual control of locomotion in Parkinson's disease. *Brain* **122**(1):111–120.

Badarny, S., Aharon-Peretz, J., Susel, Z., Habib, G. and Baram, Y. (2014). Virtual reality feedback cues for improvement of gait in patients with Parkinsons disease. *TOHM* **4**:225.

Bagley, S., Kelly, B., Tunnicliffe, N., Turnbull, G. I. and Walker, J. M. (1991). The effect of visual cues on the gait of independently mobile Parkinson's disease patients. *Physiotherapy* **77**:415–420.

Baram, Y. (1976). Information, consistent estimation and dynamic system identification. PhD thesis, Report ESL-R-718, Electronic System Laboratory, Massachusetts Institute of Technology, Cambridge.

Baram, Y. (1981). Distance measures for stochastic models. *Int J Contr* **33**(1):149–157.

Baram, Y. (1990). Construction of linear predictors for stationary vector sequences. *IEEE Trans Automat Contr* **35**(2):236–239.

Baram, Y. (1996). A bird's eye view on the descent trajectory. *IEEE Trans Aerospace Electr Syst* **32**(3):1085–1087.

Baram, Y. (1999). Walking on tiles. *Neur Proc Lett* **10**:81–87.

Baram, Y. (2004). Closed-loop augmented reality apparatus. US Patent No. 6,734,834-B1.

Baram, Y. (2021). *The Subcritical Brain: A Sinergy of Neural Circuits in Memory, Cognition and Sensorymotor Control.* World Scientific Publishing, Singapore.

Baram, Y. and Barniv, Y. (1996). Obstacle detection by learning binary expansion patterns. *IEEE Trans Aero Electr Sys.* **32**(1):191–198.

Baram, Y. and Be'eri, Y. (1981). Stochastic model simplification. *IEEE Trans Automat Contr* **26**(2):379–390.

Baram, Y. and Eidelman, D. (1985). An information approach to fixed gain design. *IEEE Trans Aerospace Electr Syst* **21**(1):47–55.

Baram, Y. and Kailath, T. (1988). Estimability and regulability of linear systems. *IEEE Trans Automat Contr* **33**(12):1116–1121.

Baram, Y. and Lenger, R. (2012). Gait improvement in patients with cerebral palsy by visual and auditory feedback. *Neuromodulation* **15**(1):48–52.

Baram, Y. and Miller, A. (2006). Effects of virtual reality cues on gait in multiple sclerosis patients. *Neurology* **66**:178–181.

Baram, Y. and Miller, A. (2007). Auditory feedback for improvement of gait in multiple sclerosis patients. *J Neurol Sci* **254**:90–94.

Baram, Y. and Miller, A. (2010). Glide-Symmetric locomotion reinforcement in patients with multiple sclerosis by visual feedback. *Disabil Rehabilitation Assist Technol* **5**(5):323–326.

Baram, B. and Porat, B. (1988). Identification of minimal order state space models from input-output data. *SIAM J Control Optim* **26**(1): 56–65.

Baram, Y. and Sandell, N. R. Jr. (1978). An information theoretic approach to dynamical systems modeling and identification. *IEEE Trans Automat Contr* **23**(1):61–66.

Baram, Y. and Shaked, U. (1984). On the order of the minimal output representation of stochastic linear systems. *Int J Contr* **40**(1):179–192.

Baram, Y. and Shaked, U. (1985). Minimal order estimation of multivariable continuous-time stochastic linear systems. *IEEE Trans Automat Contr* **30**:483–484.

Baram, Y. and Shaked, U. (1986). Minimal order estimation of multivariable discrete-time stochastic linear systems. *SIAM J Control Optim* **24**(4):817–820.

Baram, Y. and Shaked, U. (1988). On the internal structure of minimal realizations. *Automatica* **24**(5):715–717.

Baram, Y., Barniv, Y. and Sony, T. (1997). Detecting collision from gray-level expansion by a neural network. *Neurocomputing* **16**:77–84.

Baram, Y., Aharon-Peretz, J. and Lenger, R. (2010). Virtual reality feedback for gait improvement in patients with idiopathic senile gait disorders and in patient with history of strokes. *J Am Geriatr Soc* **58**(1):191–192.

Baram, Y., Aharon-Peretz, J., Simionovici, Y. and Ron. L. (2002). Walking on virtual tiles. *Neural Process Lett* **16**:227–233.

Baram, Y., Aharon-Peretz, J., Badarny, S., Susel, Z. and Schlesinger, I. (2016). Closed-loop auditory feedback for the improvement of gait in patients with Parkinson's disease. *J Neurol Sci* **363**(15):104–106.

Bauer, H. (1972). *Probability Theory and Elements of Measure Theory.* Holt, Rinehart ans Winston, Inc.

Beck, A. T., Ward, C. H., Mendelson, M., Mock, J. and Erbaugh J. (1961). An inventory for measuring depression. *Arch Gen Psychiatry* **4**(6):561–571.

Behrman, A. L., Teitelbaum, P. and Cauraugh, J. H. (1998). Verbal instructional sets to normalize the temporal and spatial gait variables in Parkinson's disease. *J Neurol Neurosurg Psychiatry* **65**:580–582.

Benedetti, M. G., Piperno, R., Bonato, P., Tonini, A. and Giannini, S. (1999). Gait abnormalities in minimally impaired multiple sclerosis patients. *Mult Scler* **5**:363–368.

Benjamin, J., Li., L., Patterson, C., Greenberg, B. D., Murphy, D. L. and Hame, D. H. (1996). Population and familial association between the D4 dopamine receptor gene and measures of novelty seeking. *Nature Genet* **12**:81–84.

Berardelli, A., Rothwell, J. C., Thompson, P. D. and Hallett, M. (2001). Pathophysiology of bradykinesia in Parkinson's disease. *Brain* **124**(11): 2131–2146.

Berger, W. (1998). Characteristics of locomotor control in children with cerebral palsy. *Neurosci Biobehav Rev* **22**:579–582.

Berger, W., Quintern, J. and Dietz, V. (1982). Pathiophysiology of gait in children with cerebral palsy. *Electroencephalogr Clin Neurophysiol* **53**:538–548.

Bernhard, P. and Cohen, G. (1974). Complete dependence of the linear quadratic optimal control problem. *Ricerche di Automatic* **5**:1–15.

Bernoulli, J. (1713) *Ars Conjectandi: Usum & Applicationem Praecedentis Doctrinae in Civilibus, Moralibus & Oeconomicis*, Chapter 4 (Translated into English by Oscar Sheynin).

Bernstein, D. S. and Hyland, D. C. (1987). The optimal projection equations for reduced-order, discrete-time state estimation for linear systems with multiplicative white noise. *Syst Contr Lett* **8**(4):381–388.

Bhattacharyya, A. (1943). On a measure of divergence between two statistical populations defined by their probability distributions. *News Bull Calcutta Math Soc* **35**:99–109.

Billingsley, P. (1961) Statistical methods in Markov chains. *Ann Math Stat* **32**:12–40.

Birkhoff, G. D. (1931). Proof of the ergodic theorem. *Proc Nat Acad Sci USA* **17**(12):656–660.

Brotchie, P., Iansek, R. and Horne, M.K. (1991). Motor function of the monkey Globus pallidus. *Brain* **14**:1685–1702.

Bryson, A. E. Jr. (1978). Kalman filter divergence and aircraft motion estimators. *AIAA J Guidance Contr* **1**:71–79.

Bryson, A. E. and Johansen, D. E. (1965). Linear filtering for time-varying systems using measurements containing colored noise. *IEEE Trans Automat Contr* **10**(1):4–10.

Caglar, A. T., Gurses, H. N., Mutluay, F. K. and Kiziltan, G. (2005). Effects of home exercises on motor performance in patients with Parkinson's disease. *Clin Rehabil* **19**(8):870–877.

Caines, P. E. (1975a) A note on the consistency of maximum likelihood estimates for finite families of stochastic processes. *Ann Stat* **3**(2):539–546.

Caines, P. E. (1975b). Prediction error identification methods for stationary stochastic processes. *IEEE Trans Automat Contr* **21**(4):500–503.

Caines, P.E. (1988). *Linear Stochastic Systems*. John Wiley, New York.

Caines, P. E. and Rissanen, J. (1974). Maximum likelihood estimation of parameters in multivariate Gaussian stochastic processes. *IFFF Trans Info Theor* **20**(1):102104.

Chacon, R. V. and Ornstein, W. (1960). A general ergodic theorem. *J Math* **4**:153160.

Chen, H. F. (1985). *Recursive Estimation and Control for Stochastic Systems.* Wiley, New York.

Chen, Y.-P., Kang, T.-Y., Doong, J.-L., Shwn-Jan Lee, S.-.J., Tsai, M.-W., Jeng, S.-F. and Sung, W.-H. (2007). Use of virtual reality to improve upper-extremity control in children with cerebral palsy: A single-subject design research report. *Phys Ther* **87**:1441–1457.

Choundhary, B. (1983). *The Elements of Complex Analysis* (2nd Edition). New Age International (P) Limited, New Delhi.

Chow, J. C. (1972a). On estimating the orders of an autoregressive moving-average process with uncertain observations, *IEEE Trans Automat Contr* **17**:707–709.

Chow, J. C. (1972b). On the estimation of the order of a moving average process, *IEEE Trans Automat Contr* **17**:386–387.

Chung, K. L. (1974). *A Course in Probability Theory.* Academic Press, New York.

Cifelli, A. and Matthews, P. M. (2002). Cerebral plasticity in multiple sclerosis: Insights from fMRI. *Mult Scler* **8**(3):193–199.

Cloninger, C. R. (1987). A systematic method for clinical description and classification of personality variants. A proposal. *Arch Gen Psychiatry* **44**(6):573–588.

Cohen, R. G., Chao, A., Nutt, J. G. and Horak, F. B. (2011). Freezing of gait is associated with a mismatch between motor imagery and motor execution in narrow doorways, not with failure to judge doorway passability. *Neuropsychologia* **49**(14):3981–3988.

Colborne, G. R., Wright. F.V. and Naumann, S. (1994). Feedback of striceps surae EMG in gait in children with cerebral palsy. *Arch Phys Med Rehabil* **75**:40–45.

Conrad, L. and Bleck, E. E. (1980). Augmented auditory feedback in the treatment of equinus gait in children. *Dev Med Child Neurol* **22**:713–718.

Côté, L. and Crutcher, M. D. (1991). The basal ganglia. *Principl Neur Sci* **42**:647–659.

Cramer, H. (1946). *Mathematical Methods of Statistics.* Princeton University Press.

Cubo, E., Leurgans, S. and Goetz, C. G. (2004). Short-term and practice effects of metronome pacing in Parkinson's disease patients with gait freezing while in the 'on' state: Randomized single blind evaluation. *Parkinsonism Relat Disord* **10**(8):507–510.

Davies, M. N. O. and Green, P. R. (1990). Optic flow variables trigger landing in hawk but not in pigeons. *Naturwissenschaften* **77**:142–144.

Davison, E. J. (1966). A method for simplifying linear dynamic systems. *IEEE Trans Automat Contr* **11**(1):93–101.

Del Olmo, F. and Cuderio, J. (2003). A simple procedure using auditory stimuli to improve movement in Parkinson's disease: A pilot study. *Neurol Clin Neurophys* **2**:1–15.

Delval, A., Moreau, C., Bleuse, S., Tard, C., Ryckewaert, G., Devos, D. and Defebvre, L. (2014). Auditory cueing of gait initiation in Parkinson's disease patients with freezing of gait. *Clin Neurophysiol* **125**(8):1675–1681.

Dem'yanov, V. F. and Malozemav, V. N. (1974). *Introduction to Minimax*. Wiley, New York.

Djamshidian, A., O'Sullivan, S. S., Wittmann, B. C., Lees, A. J. and Averbeck, B. B. (2011). Novelty seeking behaviour in Parkinson's disease. *Neuropsychologia* **49**(9):2483–2488.

Doob, J. L. (1934). Probability and statistics. *Trans Amer Math Soc* **36**(1934)759–775.

Doob, J. L. (1953). *Stochastic Processes*. John Wiley and Sons, Inc., New York.

Dozza, M., Chiari, L, Chan, B., Rocchi, L., Horak, F. B. and Cappello, A. (2005). Influence of a portable audio-biofeedback on structural properties of postural sway. *J Neuroeng Rehab* **2**:13–28.

Dunne, J. W., Hankey, G. J. and Edis, R. H. (1987). Parkinsonism: Upturned walking stick as an aid to locomotion. *Arch Phys Med Rehabil* **68**:380–381.

Dursun, E., Dursun, N. and Alican. D. (2004). Effects of biofeedback treatment on gait in children with cerebral palsy. *Disabil Rehabil* **26**:116–120.

Ebstein, R. P., Novick, O., Umansky, R., Priel, B., Osher, Y., Blaine, D., Bennett, E. R., Nemanov, L., Katz, M. and Belmaker, R. H. (1996). Dopamine D4 receptor (D4DR) exon III polymorphism associated with the human personality trait of novelty seeking. *Nature Genet* **12**: 78–80.

Edgeworth, F. Y. (1908). On the probable error of frequency constants. *J Roy Stat Soc* **71**(2):381–397.

Espay, A. J., Li, J. Y., Johnston, L., Chen, R. and Lang, A. E. (2005). Mirror movements in parkinsonism: Evaluation of a new clinical sign. *J Neurol Neurosurg Psychiatr* **76**(10):1355–1358.

Espay, A. J., Baram, Y., Dwivedi, A. K., Shukla, R., Gartner, M., Gaines, L., Duker, A.P. and Revilla, F. J. (2010). At-home training with closed-loop augmented-reality cueing device for improvement of gait in patients with Parkinson's disease. *J Rehabil Res Dev* **47**(6):573–582.

Etkin, B. (1972). *Dynamics of Atmospheric Flight*. Wiley, New York.

Fahn S. and Elton, R. L. (1989). Unified Parkinson's Disease Rating Scale. In: *Recent Developments in Parkinson's Disease*, Fahn, S., editor. Raven Press, New York, 1987, pp. 153–163.

Fauchard-Renard, C., Renard, J. F., Miret, N., Hannequin, D., Mihout, B. and Weber, J. (2001). Therapeutic efficacy during active phases of multiple sclerosis: Gait analysis and comparison with EDSS score. *Rev Neurol (Paris)* **157**:649–654.

Faurre, P. I. (1969). Identification par minimization d'une representation markovianne de processus aleatoire. *Symp Optimiz*, vol. 132.

Faurre, P. I. (1976). Stochastic realization algorithms. In: *System Identification*, Mehra R. K. and Lainiotis, D. G., editors. Academic, New York.

Fine, T. L. and Hwang, W. G. (1979). Consistent estimation of system order. *IEEE Trans Automat Contr* **24**:387–402.

Fisher, R. A. (1922). On the mathematical foundation of theoretical statistics. *Phil Trans Royal Soc London A* **222**:309–368.

Fletcher, R. (1979). A new approach to variable metric algorithm. *IEEE Trans Automat Contr* **24**(13):317–322.

Fletcher, W. A. and Aharpe, J. A. (1988). Smooth pursuit dysfunction in Alzheimer's disease. *Neurology* **38**(2):272.

Frankel, T. (2012). *The Geometry of Physics: An Introduction* (3rd Edition). Cambridge University Press.

Gauss, C. F. (1809). *Theoria Motus Corporium Coeletium In Sectionibus Conicis Solem Ambientium*. Hamburg.

Gelb, A., editor. (1974). *Applied Optimal Estimation*. MIT Press, Cambridge.

Genesio, R. and Milanese M. (1976). A note on the derivation and use of reduced order models. *IEEE Trans Automat Contr* **21**(1):118–122.

Gevers, M. R. and Kailath, T. (1973). An innovations approach to least-squares estimation-part VI: Discrete-time innovations representations and recursive estimation. *IEEE Trans Automat Contr* **18**(6):588–600.

Giladi, N., Shabtai, H., Simon, E. S., Biran, S., Tal. J. and Korczyn, A. D. (2000). Construction of freezing of gait questionnaire for patients with Parkinsonism. *Parkinsonism Relat Disord* **6**(3):165–170.

Gilbert, E. G. (1963). Controllability and observability in multivariable control systems. *J SIAM A Contr*, pp. 28–151.

Glover, K. and Willems, J. C. (1974). Parametrization of linear dynamical systems: Canonical forms and identification, *IEEE Trans Automat Contr* **19**(6):640–646.

Golub, G. and Kahan, W. (1965). Calculating the singular values and pseudo-inverse of a matrix. *Siam J Numer Anal* **2**(2):205–224.

Goodrich, R. and Caines, P. (1979). Necessary and sufficient conditions for local identifiability. *IEEE Trans Automat Contr* **24**(1):125–127.

Graef, S., Biele, G., Krugel, L. K., Marzinzik, F., Wahl, M., Wotka, J., Klostermann, F. and Heekeren, H. R. (2010) Differential influence of Levodopa on reward-based learning in Parkinson's disease. *Front Hum Neurosci* **4**:169.

Gray, P. and Hildebrand, K. (2000). Fall risk factors in Parkinson's disease. *J Neurosci Nurs* **32**(4):222–228.

Griffin, H. J., Greenlaw, R., Limousin, P., Bhatia, K., Quinn, N. P. and Jahanshahi, M. (2011). The effect of real and virtual visual cues on walking in Parkinson's disease. *J Neurol* **258**(6):991–1000.

Grossberg, S. (1993). A solution of the figure-ground problem for biological vision. *Neural Networks* **6**(4):463–483.

Grossberg, S., Mingolla, E. and Todorovoc, D. (1989). A neural network architecture for preattentive vision. *IEEE Trans Biomed Eng* **36**(1):65–84.

Gupta, N. K. and Mehra, R. K. (1974). Computational aspects of maximum likelihood estimation and reduction in sensitivity function calculation. *IEEE Trans Automat Contr* **19**:774–783.

Halmos, P. R. (1956). *Lectures on Ergodic Theory.* Chelsea Publication Co., New York.

Hanakawa, T., Fukuyama, H., Katsumi, Y., Honda, M. and Shibasaki, H. (1999). Enhanced lateral pre-motor activity during paradoxical gait in Parkinson's disease. *Ann Neurol* **45**(3):329–336.

Hannan, E. J. (1970). *Multiple Time Series.* John Wiley, New York.

Hawkes, R. M. and Moore, J. B. (1976). Performance bounds for adaptive estimation. *Proc IEEE***64**(8):1143–1150.

Helmich. R. C., De Lange, F. P., Bloem, B. R, and Toni, I. (2007). Cerebral compensation during motor imagery in Parkinson's disease. *Neuropsychologia* **45**(10):2201–2215.

Hespanha, J. (2018). *Linear Systems Theory* (2nd Edition). Princeton University Press.

Ho, B. L. and Kalman, R. E. (1966). Effective construction of linear state-variable models from input/output functions. *Regelungstechnik* **12**:545–548.

Hocherman, S. and Aharon-Peretz, J. (1994). Two-dimensional tracing and tracking in patients with Parkinson's disease. *Neurology* **44**(1):111–116.

Hoehn, M. M. and Yahr, M. D. (1967). Parkinsonism: Onset, progression, and mortality. *Neurology* **17**(5):427–442.

Hooper, M. J. (1978). *Harwell Subroutine Library Catalogue.* Computer Science and System Division, Atomic Energy Research Establishment. Harwell, Oxfordshire, England.

Howe, T. E., Lovgreen, B., Cody, F. W. J., Ashton V. J. and Oldham, J. A. (2003). Auditory cues can modify the gait of persons with early-stage

Parkinson's disease: A method for enhancing Parkinsonian walking performance? *Clin Rehab* **17**:363–367.

Hutton, J. T., Nagel, J. A. and Loewenson, R. B. (1984). Eye tracking dysfunction in Alzheimer-type dementia. *Neurology* **34**(1):99–102.

Ikemoto, S. (2007). Dopamine reward circuitry: Two projection systems from the ventral midbrain to the nucleus accumbens-olfactory tubercle complex. *Brain Res Rev* **56**(1):27–78.

Imhoff, J. P. (1961). Computing the distribution of quadratic forms in normal variables. *Biometrika* **48**:419.

Jazwinsky, A. H. (1970). *Stochastic Processes and Filtering Theory.* New York: Academic Press.

Jeffreys, H. (1946). An invariant form for the prior probability in estimation problems. *Proc Royal Soc London A* **186**:453–461.

Jensen, J. L. W. V. (1906). Sur les fonctions convexes et les inégalités entre les valeurs moyennes. *Acta Mathematica* **30**(1):175–193.

Jiang, Y. and Norman, K. E. (2006). Effects of visual and auditory cues on gait initiation in people with Parkinson's disease. *Clin Rehabil* **20**(1):36–45.

Johnson, N. L. and Kotz, S. (1970). *Continuous Univariate Distribution, Part II.* John Wiley, New York.

Joyce, J. (2003). Bayes' theorem. In: *The Stanford Encyclopedia of Philosophy,* Edward, N., editor.

Kailath, T. (1967). The divergence and the Bhattacharyya distance measure in signal selection. *IEEE Trans Comm Tech* **15**(1):52–60.

Kailath, T. (1979). *Linear Systems.* Prentice-Hall, Englewood Cliffs, New Jersey.

Kalman, R. E. (1960a). On the general theory of control systems. *Proceedings of the First International Congress on Automatic Control,* pp. 481–491.

Kalman, R. E. (1960b). A new approach to linear filtering and prediction problems. *J Basic Eng Trans,* pp. 35–45.

Kalman, R. E. (1963). Mathematical description of linear dynamical systems. *SIAM Industr Appl Math A Contr,* pp. 152–192.

Kassam, S. A. and Lim, T. L. (1977). Robust Wiener filters. *J Franklin Inst.* **304**:171–185.

Kendall, M. G. and Stuart, A. (1973). *The Advanced Theory of Statistics, Vol. 2.* Griffin, London.

Ketalaer, P. and De Wolf, L. (2000). New technologies in MS: Applications of virtual reality. *Int MSJ* **8**:11–19.

Klockgether, T., Borutta, M., Rapp, H., Spieker, S. and Dichgans, J. (1995). A defect of kinesthesia in Parkinson's disease. *Mov Disord* **10**(4):460–465.

Koenderink, J. J. and van Doorn, A. J. (1975). Invariant properties of the motion parallax field due to the movement of rigid bodies relative to an observer. *Optica Acta* **22**(9):773–791.

Kompoliti. K, Goetz, C. G., Leurgans, S., Morrissey, M. and Siegel, I. M. (2000). On freezing in Parkinson's disease: Resistance to visual cue walking devices. *Mov Disord* **15**(2):309–312.

Kritikos, A., Leahy, C., Bradshaw, J. L., Iansek, R., Phillips, J. G. and Bradshaw, J. A. (1995). Contingent and non-contingent auditory cueing in Parkinson's disease. *Neuropsych* **33**(10):1193–1203.

Kullback, S. (1959). *Information Theory and Statistics.* John Wiley and Sons, Inc., New York.

Kullback, S. and Leibler, R. A. (1951). On information and sufficiency. *Ann Math Stat* **12**:79–86.

Kung, S. Y. (1977). Multivariable and multidimensional systems: Analysis and design, Ph.D. dissertation, Stanford University, Stanford, CA.

Kuo, B. C. (1981). *Automatic Control Systems.* Prentice-Hall, Englewood Cliffs, New Jersey.

Kurtzke, J. F. (1993). Rating neurological impairment in multiple sclerosis: An (EDSS). *Neurology* **33**:1444–1452.

Kwakernaak, H. and Sivan, R. (1972). *Linear Optimal Control Systems.* Wiley, New York.

Laplace, P. S. (1820). *Theorie Analytique de Probabilities* (3rd Edition). Paris, pp. 309–354.

Larimore, W. E. (1983). System identification, reduced order filtering and modelling via canonical variates analysis. *Proc Am Contr Conf,* Rao, H. S. and Dorato, T., editors. IEEE, New York, pp. 445–451.

Lecam, L. (1953). On some asymptotic properties of maximim likelihood estimates and related Bayes' estimates. *Univ Calif Publ Statistics* **1**:277–330.

Ledebt, A., Becher, J., Kapper, J., Rozenbdaalr, Bakker, R., Leenders, I. C. and Savelsbergh, G. J. P. (2005). Balance training with visual feedback in children with hemiplegic cerebral palsy: Effect on stance and gait. *Motor Control* **9**(4):459–468.

Lee, D. N., Davies, M. N. O., Green, P. R. and Van Der Weel, F. R. (Ruud). (1993). Visual control of velocity of approach by pigeons when landing. *J Exp Biol* **180**(1):85–104.

Lewis, G. N, Byblow, W. D. and Walt, S. E. (2000). Stride length regulation in Parkinson's disease: The use of extrinsic, visual cues. *Brain* **123**:2077–2090.

Liporace, L. A. (1971). Variance of Bayes estimates. *IEEE Trans Info Theor* **17**(6):665–669.

Liu, A. R. and Bitmead, R. R. (2011). Stochastic observability in network state estimation and control. *Automatica* **47**:65–78.

Livio, M. (2005). *The Equation that Couldn't be Solved.* Simon & Schuster, New York.

Ljung, L. (1974a). Convergence of recursive stochastic algorithms. Report 7403, Division of Automatic Control at Lund Institute of Technology.

Ljung, L. (1974b). On consistency for prediction error identification methods. Report 7405, Division of Automatic Control at Lund Institute of Technology

Ljung, L. (1975). On consistency and identifiability. Report 7521, Division of Automatic Control at Lund Institute of Technology.

Looze, D., Poor, H., Vastola, K. and Darragh, J. (1983). Minimax control of linear stochastic systems with noise uncertainty. *IEEE Trans Automat Contr* **28**(9):882–888.

Lord, S. E., Wade, D. T. and Halligan, P. W. (1998). A comparison of two physiotherapy treatment approaches to improve walking in multiple sclerosis: A pilot randomized controlled study. *Clin Rehabil* **12**:477–486.

Mahalanabis, A. K. and Ray, G. (1986). Modelling large power systems for efficient load frequency control. *Math Modelling* **7**:259–272.

Martin, J. P. (1967). Locomotion and the basal ganglia. In: *The Basal Ganglia and Posture*, Martin, J. P., editor. Pitman Medical, London.

Maurer, C., Mergner, T. and Peterka, R. J. (2003). A multisensory posture control model of human upright stance. *Prog Brain Res* **142**:189–201.

Maxwell, J. C. (1868). *On Governors.* Proceedings of the Royal Society.

McIntosh, G. C., Brown, S. H., Rice, R. R. and Thaut, M. H. (1997). Rhythmic auditory-motor facilitation of gait patterns in patients with Parkinson's disease. *J Neurol Neurosurg Psych* **62**:22–26.

Menza, M. A., Golbe, L. I., Cody, R. A. and Forman N. E. (1993) Dopamine-related personality traits in Parkinson's disease. *Neurology* **43**:505–508.

Mevellec, E., Lamotte, D., Cantalloube S., Amarenco, G. and Thoumie, P. (2003). Relationship between gait speed and strength parameters in multiple sclerosis [in French]. *Ann Readapt Med Phys* **46**:85–90.

Miendlarzewska, W. J. and Trost, E. A. (2013). How musical training affects cognitive development: Rhythm, reward and other modulating variables. *Front Neurosci* **7**:279.

Mohler, R. R. and Hwang, C. S. (1988). Nonlinear data observability and information-engineering and applied mathematics. *J Franklin Inst* **325**:443–464.

Moore, J. B. and Hawkes, R. M. (1974). Performance of Bayesian parameter estimators for linear signal models. Report EE7410, University of Newcastle, Australia.

Moore, O., Peretz, C. and Giladi, N. (2007). Freezing of gait affects quality of life of peoples with Parkinson's disease beyond its relationships with mobility and gait. *Mov Disord* **22**(15):2192–2195.

Morris, M. E, Cantwell, C., Vowels, L. and Dodd, K. (2002). Changes in gait and fatigue from morning to afternoon in people with multiple sclerosis. *J Neurol Neurosurg Psychiatry* **72**:361–365.

Morris, M. E., Iansek, R., Matyas, T. A. and Summers, J. J. (1994a). The pathogenesis of gait hypokinesia in Parkinson's disease. *Brain* **117**:1160–1181.

Morris, M. E., Iansek, R., Matyas, T. A. and Summers, J. J. (1994b). The ability to modulate walking cadence remains intact in Parkinson's disease. *J Neurol Neurosurg Psychiatry* **57**:1532–1534.

Morris M. E., Iansek, R., Matyas, T. A. and Summers, J. J. (1996). Stride length regulation in Parkinson's disease: Normalizations strategies and underlying mechanisms. *Brain* **119**:551–568.

Moustafa, A. (2010). Levodopa enhances reward learning but impairs reversal learning in Parkinson's disease patients. *Front Hum Neurosci* **4**:169.

Nataraj, R. and van den Bogert, A. J. (2017). Simulation analysis of linear quadratic regulator control of Sagittal-plane human walking: Implications for exoskeletons. *J Biomech Sci Eng* **139**(10).

Nelson, R. C. and Aloimonos, J. (1989). Obstacle avoidance using flow field divergence. *IEEE Trans Patt Anal Mach Intell* **11**(10):1102–1106.

Nieuwboer, A., Kwakkel, G., Rochester, L., Jones, D., van Wegen, E., Willems, A. M., Chavret, F., Hetherington, V., Baker, K. and Lim, I. (2007). Cueing training in the home improves gait-related mobility in Parkinson's disease. The RESCUE trial. *J Neurol Neurosurg Psychiatry* **78**:134–140.

Nikodym, O. (1930). Sur une généralisation des intégrales de M. J. Radon (PDF). *Fundamental Math* (in French) **15**:131–179.

Orsnes, G. B., Sorensen, P. S., Larsen, T. K. and Ravnborg, M. (2000). Effect of baclofen on gait in spastic MS patients. *Acta Neurol Scand* **101**:244–248.

Pacchetti, C., Aglieri, R., Mancini, F., Martignoni, E. and Nappi, G. (1998). Active music therapy and Parkinson's disease: Methods. *Funct Neurol* **13**(1):57–67.

Palliyath, S., Hallett, M., Thomas, S. L. and Lebiedowska, M. K. (1998). Gait in patients with cerebellar ataxia. *Mov Disord* **13**:958–964.

Paltsev, Y. I. and Elner, A. M. (1967). Change in the functional state of the segmental apparatus of the spinal cord under the influence of sound stimuli and its role in voluntary movement. *Biophysics* **12**:1219–1226.

Poor, H. V. (1980). On robust Wiener filtering. *IEEE Trans Automat Contr* **25**(3):531–536.

Poor, H. V. and Looze, D. P. (1981). Minimax state estimation for linear stochastic systems with noise uncertainty. *IEEE Trans Automat Contr* **26**(4):902–906.

Poser, C. M., Paty, D. W., Scheinberg L., McDonald, W. I., Davis, F. A., Ebers, G. C., Johnson. K. P., Sibley, W. A., Siberberg, D. H. and Turtellotte, W. W. (1983). New diagnostic criteria for multiple sclerosis: Guidelines for research protocols. *Ann Neurol* **13**:227–231.

Prazdny, K. (1981). Determining the instantaneous direction of motion from optical flow generated by a curvilinearly moving observer. *Comput Vis Image Underst* **17**(3):238–248.

Prothero, J. (1993). The treatment of akinesia using virtual images. M.Sc. thesis, University of Washington, College of Engineering.

Rao, C. R. (1973). *Linear Statistical Inference.* Wiley, New York.

Rendic. Circ. Mat. Palermo 22 (1906) 1–74.

Riess, T. and Weghorst, S. (1995). Augmented reality in the treatment of Parkinson's disease. *Proceedings of the 7th Medicine Meets Virtual Reality III Conference*, Amsterdam, The Netherlands.

Ringach, D. L and Baram, Y. (1992). Obstacle detection by diffusion. Technical report No. CIS–9220, Computer Science Department, Technion, Israel Institute of Technology, November.

Ringach, D. L. and Baram, Y. (1994). A diffusion mechanism for obstacle detection from size-change information. *IEEE Trans Pattern Anal Mach Intell* **16**(1):76–80.

Rissanen, J. (1971). Recursive identification of linear systems, *J Soc Industr Appl Math, A: Contr* **9**:420–430.

Rissanen, J. and Caines, P. E. (1974). Consistency of maximum likelihood estimators for ARMA processes. Control Systems Report 7424, Department of Electrical Engineering, University of Toronto.

Rissanen, J. and Ljung, L. (1975). *Mathematical Systems Theory.* Udine Springer Verlag.

Rizzo, A.A., Schulthheis, M., Kerns, K. A. and Mateer, C. (2004). Analysis of assets for virtual reality applications in neuropsychology. *Neuropsychol Rehabil* **14**:207–239.

Rochester, L., Nieuwboer, A., Baker, K., Hetherington, V., Willems, A. M., Chavret, F., Kwakkel, G., Van Wegen, E., Lim, I. and Jones, D. (2007). The attentional cost of external rhythmical cues and their impact on gait in Parkinson's disease: Effect of cue modality and task complexity. *J Neural Transm* **114**(10):1243–1248.

Rodgers, M. M., Mulcare, J. A., King, D. L., Mathews, T., Gupta, S. C. and Glaser, R. M. (1999). Gait characteristics of individuals with multiple sclerosis before and after a 6-month aerobic training program. *J Rehabil Res Develop* **36**:183–188.

Rougier, P. (1999). Influence of visual feedback on successive control mechanisms in upright quiet stance in humans assessed by fractional Brownian motion modeling. *Neurosci Lett* **142**:157–160.

Roussas, G. G. (1965). Extension to Markov processes of a result by A. Wald about the consistency of maximum likelihood estimates. *Z. Wahrscheinlichkeitstheorie* **4**:69–73.

Rudin, W. (1976). *Principles of Mathematical Analysis* (3rd Edition). McGraw-Hill, New York.

Sacks, O. W. (1973). *Awakenings.* 1st Vintage Books, New York.

Sacks, O. W. (1985). *The Man who Mistook his Wife for a Hat and Other Clinical Tales.* Summit Books, New York,

Salimpoor, V. N., Zald, D. H., Zatorre, R. J., Dagher, A. and McIntosh, A. R. (2015). Predictions and the brain: How musical sounds become rewarding. *Trends Cogn Sci* **19**(2):86–91.

Salmon, D. M. (1968). Minimax controller design. *IEEE Trans Autom Contr* **13**(4):369–376.

Schelter, B., Winterhalder, M., Eichler, M., Peifer, M., Hellwig, B., Guschlbauer, B., Lucking, H. C., Dahlhous, R. and Timmer, J. (2006). Testing for directed influences among neural signals using partial directed coherence. *J Neurosci Methods* **152**(1–2):210–219.

Schweppe, F. C. (1973). *Uncertain Dynamic Systems.* Prentice-Hall, Englewood Cliffs, New Jersey.

Seeger, B. R. and Caudrey, D. J. (1983). Biofeedback therapy to achieve symmetrical gait in hemiplegic cerebral palsy: Long-term efficacy. *Arch Phys Med Rehabil* **64**:160–162.

Seeger, B. R., Caudrey, D. J. and Scholes J. R. (1981). Biofeedback therapy to achieve symmetrical gait in hemiplegic cerebral palsied children. *Arch Phys Med Rehabil* **62**:364–368.

Shaked, U. and Karcanias, N. (1976). The use of zeros and zero directions in model reduction. *Int J Contr* **23**(1):113–135.

Shamash, Y. (1975). Linear systems reduction using Pade approximation to allow retention of dominant modes. *Int J Contr* **21**:257–272.

Sidaway, B., Anderson, J., Danielson, G., Martin, L. and Smith, G. (2006). Effects of long-term gait training using visual cues in an individual with Parkinson disease. *Phys Ther* **86**(2):186–194.

Silverman, L. M. (1971). Realization of linear dynamical systems, *IEEE Trans Automat Contr* **16**:554–567.

Silverman, L. M. (1976). Discrete Riccati equations: Alternative algorithms, asymptotic properties and system theoretic interpretation. In: *Control and Dynamic Systems*, Leondes, C. T., editor. **12**:313–386.

Sims, C. S. and Asher, R. B. (1978). Optimal and suboptimal results in full and reduced order linear filtering. *IEEE Trans Automat Contr* **23**(3):469–472.

Sinha, N. K. and Bereznai, G. T. (1971). Optimum approximation of large dynamic systems by low-order models. *Int J Contr* **14**:951–959.

Stein, G., Hartmann, G. and Hendrick, R. (1977). Adaptive control laws for F-8 flight test. *IEEE Trans Automat Contr* **22**(5):758–768.

Stout, W. F. (1974). *Almost Sure Convergence.* Academic Press, New York.

Thaut. M. H., McIntosh, G. C. and Hoenberg, V. (2014). Neurobiological foundations of neurologic music therapy, rhythmic entrainmrnt and the motor system. *Front Psychol* **5**:1185

Thaut, M. H., McIntosh, G. C., Rice, R. R., Miller, R. A., Rathbun, J. and Brault J. M. (1996). Rhythmic auditory stimulation in gait training for Parkinson's disease patients. *Move Disord* **11**:193–200.

The free dictionary. (2018). http://www.thefreedictionary.com

Thoumie, P. and Mevellec, E. (2002). Relation between walking speed and muscle strength is affected by somatosensory loss in multiple sclerosis. *J Neurol Neurosurg Psychiatry* **73**:313–315.

Tse, E. (1973). Information matrix and local identifiability of parameters. *Proceedings of the Joint Automatic Control Conference*, Paper 20–3, Columbus, Ohio.

Tse, E. (1976). Identification and estimation of dynamic linear models with unknown structure. *Proceedings of the Joint Automatic Control Conference*, Purdue University.

Tse, E. and Weinert, H. L. (1975). Structure determination and parameter identification for multivariable stochastic linear systems. *IEEE Trans Automat Contr* **20**(5):603–613.

Tugnait, J. (1982). Comment on an approximation coefficient proposed in "Stochastic Model Simplification". *IEEE Trans Automat Contr* **27**(4):101–104.

van der Heide, J. C. and Hadders-Algra, M. (2005). Postural muscle dyscoordination in children with cerebral palsy. *Neural Plast* **12**:197–203.

van Praag, H., Kempermann, G. and Gage, F. H. (1999a). Running increases cell proliferation and neurogenesis in the adult mouse dentate gyrus. *Nat Neurosci* **2**(3):266–270.

van Praag, H., Christie, B. R., Sejnowski, T. J. and Gage, F.H. (1999b). Running enhances neurogenesis, learning, and long-term potentiation in mice. *Proc Nat Acad Sci USA* **96**(23):13427–13431.

Van Wegen, E., De Goede, C., Lim, I., Rietberg, M., Nieuwboer, A., Willems, A., Jones, D., Rochester, L., Hetherington, V., Berendse, H., Zijlmans, J., Wolters, E. and Kwakkel, G. (2006). The effect of rhythmic somatosensory cueing on gait in patients with Parkinson's disease. *J Neurol Sci* **248**(1–2):210–214.

Vasilakos, K. and Beuter, A. (1993). Effects of noise on a delayed visual feedback system. *J Theor Biol* **165**(3):389–407.

Velu, P. D., Mullen, T., Noh, E., Valdivia, M., Poizner, H., Baram, Y. and de Sa, V. R. (2013). Effect of visual feedback on the occipito-parietal-motor network in Parkinson's disease patients with freezing of gait. *Front Neurol* **4**:209.

Vieregge, P., Stolze, H., Klein, C. and Heberlein I. (1997). Gait quantification in Parkinson's disease-Locomotor disability and correlation to clinical rating scales. *J Neural Transm* **104**(2–3):237–248.

Visser, H. (1983). Gait and balance in senile dementia of Alzheimer's type. *Aging* **12**(4):296–301.

Wald, A. (1949). Note on the consistency of the maximum likelihood estimate. *Ann Math Stat* **20**:595–601.

Wassum, K. M., Ostlund, S. B., Balleine, B. W. and Maidment, N. T. (2011). Differential dependence of Pavlovian incentive motivation and instrumental incentive learning processes on dopamine signaling. *Learn Mem* **18**(7):475–483.

Webster, K. E., Wittwer, J. E. and Feller, J. A. (2005). Validity of the GAITRite walkway system for the measurement of averaged and individual step parameters of gait. *Gait Posture* **22**(4):317–321.

Weghorst, S., Prothero, J. and Furness, T. (1994). Virtual images in the treatment of Parkinson's disease akinesia. *Proceedings of the 7th Medicine Meets Virtual Reality II Conference*, pp. 242–243.

Weiner, N. (1949). *Extrapolation, Interpolation and Smoothing of Stationary Time Series.* MIT Press, Cambridge.

Woodside, C. M. (1970). Estimation of the order of linear systems, *IFAC Symposium of Identification and System Parameter Estimation.* Prague.

Woodward, P. M. (1980). *Probability and Information Theory with Applications to Radar.* Artech House, Norwood, MA.

Woollacott, M. H. and Burtner, P. (1996). Neural and musculoskeletal contribution to the development of stance balance control in typical children and in children with cerebral palsy. *Acta Paediatr Suppl* **416**:58–62.

Zaccara, G., Gangemi, P. F., Muscas, G.C., Paganini, M., Pallanti, S., Parigi, A., Messori, A. and Arnetoli, G. (1992). Smooth-pursuit eye movements: Alterations in Alzheimer's disease. *Neurol Sci* **112**(1–2):81–89.

Zolotarev, V. M. (1977). Ideal metrics in the problem of approximating the distributions of sums of independent random variables. *Teor Veroyatnost i Primenen* **22**(3):449–465.

Index

www.ingramcontent.com/pod-product-compliance
Lightning Source LLC
LaVergne TN
LVHW010331220926
848484LV00004BB/877